Praise for the stories in
Wild Thing, *I Think I Love You!*

"King's great gift is her ability to transport her readers to wherever she happens to be, and to share her experience. While she's tracking lemurs through Madagascar's rain forests, for instance, we viscerally feel the pulsing ... humidity of the leech-infested environment, even as we become deeply concerned about the plight of those charismatic yet fragile remote human relatives."

—Dr. Ian Tattersall, Author, *Masters of the Planet: The Search for Our Human Origins*
Curator Emeritus, Division of Anthropology
American Museum of Natural History

HUMAN AUTHORED

"King's powerful piece on saving wild elephants is a moment you should not miss—even if it hurts."

—Natalie Lefevere, Editor
Ethical Traveler

"The Bald Eagles of Klamath Basin have been called the greatest gathering of eagles outside of Alaska. And King has done the hard work for you—rousing before dawn, wearing five layers of insulation to fend off the icy air, repeatedly wiping the fog off her binoculars. And then there they are—massive dark-bodied eagles flying out of a shallow canyon. The best thing about King's essays is that you already know Laurie. She is your pal, your sister, your shadow-person who does the adventure with you and says just the thing you wish you'd said."

—Allen Fish, former Director,
Golden Gate Raptor Observatory

"Talk about "out there" ... from twenty-foot-long earthworms to the perils of collecting rhinoceros urine, King uncovers the wild and wacky side of wildlife conservation. I laughed. I learned. I LOVED it!

—Linda Watanabe McFerrin, Author
Navigating the Divide

"Whether she's describing a thrilling adventure with massive beasts of the jungle or experiencing a moment of awe in a butterfly sanctuary, the work is captivating... entertaining...Engaging and packed with countless intriguing factoids about the natural world."

—*Kirkus Reviews*
"Our Verdict: GET IT"

Wild Thing, *I Think I Love You!*

Close Encounters with
Wildlife Around the World

Wild Thing,
I Think I Love You!

Close Encounters with Wildlife Around the World

Laurie McAndish King

Destination Insights

Similar versions of most of the essays in this collection appeared previously, primarily in the author's earlier books, with a small number published elsewhere or appearing here for the first time. In the interest of privacy, some names and the identifying details about individuals and exact locations have been changed.

Photos are courtesy of the author except as noted: Cover and page 8, David Groves / Wikimedia Commons; page xviii, Paul Danese / Wikimedia Commons; page 34, Ilham.nurwansah / Wikimedia Commons; page 50, Ltshears / Wikimedia Commons; page 60, Charles J. Sharp / Wikimedia Commons; page 84, Flickr by Naparazzi / Wikimedia Commons; page 106, News Ltd/Newspix (Australia); page 160, a photo of a photograph on display at the Asa Wright Nature Center; page 172, Diego Delso / Wikimedia Commons; page 182, Creative Commons: Julia's Travels; pages 190 and 200, James M. Shubin.

Published by: Destination Insights
Cover design, interior design by JM Shubin

CATALOGING IN PUBLICATION DATA:
Wild Thing, I Think I Love You! by Laurie McAndish King

ISBN 978-1-7327747-6-6

HUMAN AUTHORED Reg: 7850306,https://authorsguild.org/human

Dedicated to the memory
of Dr. Jane Goodall,
who has inspired
generations of conservationists

Contents

Unique Species

Life on the Edge

AWESTRUCK AND INSPIRED

On the Edge of Wild

My work as a travel writer has allowed me to observe nature in a way that would be impossible at home. The awe-inspiring African savannah, for example, is the best place in the world for viewing large, charismatic animals: lazing lions, grazing giraffe, and zebras with their camouflage stripes. Some of the world's best bird-watching is on the Caribbean island of Trinidad. Madagascar is the *only* place to see lemurs in their native habitat.

Visiting ecosystems around the world brought me into close contact with species I began to love and wanted to help protect. Each encounter also raised questions about the ways we humans interact with wildlife—some helpful, some harmful, some seemingly out of our control:

- Tracking critically endangered lemurs in Madagascar got me thinking about the inherent conflict between humans and animals in a land where they both need the same space to feed themselves.

- Visiting an elephant orphanage gave me an immense love and appreciation for pachyderms, and taught me about some of the more subtle problems habitat loss creates.

- Watching a leatherback turtle laying her eggs was an empathetic moment I could never have anticipated, and made me care a lot more about the light pollution that affects the turtles' survival.

These stories take place where human and animal lives intersect—on the edge of wild. They are an invitation to notice the lives of wild animals, to step a little closer, to care even more deeply. Some provide solutions, others offer hope; still other stories may foster determination to share Earth more equitably. Every one was written as a way to honor and rekindle our awe, respect, and deep love of nature.

A Little Too Close

A red-tailed hawk, whose long halux claw can measure up to 1.3 inches

Hooked on Hawk Hill

Captured by a raptor in Sausalito, California

When I flinched even the slightest bit, the red-tailed hawk tightened her vice-like grip on my right thigh. Her talon was more than an inch long: sturdy, with a finely pointed tip that had slipped effortlessly through my jeans and pierced my soft flesh, unleashing excruciating pain. The hawk was large—her wingspan more than four feet. I figured I needed at least four hands to free myself: one to hold the bird, whose instinct was to grasp me and fly away; one to hold her free foot, so she didn't seize me with that; and two to hold the offending foot and pull out the talon, which was curved like a huge fish hook. It was quite a predicament, and not an uncommon one among inexperienced raptor banders.

Actually, as soon as I held my first bird I was hooked—in the metaphorical sense. There is something magical about seeing a bird of prey up close, looking it in the eye, and feeling its raw power. There is something addicting about being a witness to its magnificent wildness.

That's why I volunteered to be a bander with the Golden Gate Raptor Observatory (GGRO), a citizen-science organization that's been tracking birds of prey in northern California for more than thirty years. It's why I was in a small, dark bird blind on Hawk Hill, in the windswept headlands just north of San Francisco. And it's why I was stuck, immobile, with a sharp talon in my leg and searing pain on my mind.

Fortunately, I was not alone. It takes a team to catch a hawk, and I was part of a group of ten banders working in four blinds that day. My blind-mate was Buzz, an experienced bander and the GGRO's Director of Research. There's no one I'd rather be in a blind with. Buzz understands the way hawks behave. He knows how to read the wind and figure out how the birds will be flying. He can look at a hawk's plumage—the subtle variations in color, the length of the flight feathers in relation to each other, the amount of wear on the feathers' edges—and determine the bird's age. And he knows how to trap a bird and handle it safely—which, clearly, I did not.

It isn't easy to catch a hawk. The other banders and I went through an extensive training program that included learning to set up three kinds of nets—bow nets, mist nets, and dho-gazzas—because different nets work best in different weather conditions and for catching different-sized birds. Nineteen species of raptors pass through the headlands; big red-tails and the smaller Cooper's hawks and sharp-shinned hawks are most common.

We use bow nets for the larger birds, and mist nets or dho-gazzas for the smaller ones—unless it's very windy, in which case we skip the dho-gazzas because they won't stay up in more than a light breeze. No matter which kind we use, catching the birds is a challenge. The altitude and direction they're flying in from, the angle of approach, the wind currents, the time of day, and the position of the sun—all these factors play into the trapping equation.

Getting the birds out of the nets is often challenging, too. Sometimes they dive deeply into the mesh, requiring us to snip the netting to disentangle them—necessitating a net-repair job that takes time away from trapping. Other times they get a wing stuck, or bite us as we try to extricate them.

We don't hold on to the hawks for very long. Once we've untangled them from the net, our job is to perform a quick set of assessments and release the birds unharmed. The first thing we do is attach a small, numbered band to the "ankle" of each bird we catch, so it can be identified if it's captured again, either here or somewhere else. Then we measure and weigh the birds and record data: each bird's age and sex; whether its crop is full or empty; whether it appears to have any injuries or abnormalities; and the length of its wing, tail, beak, and talon.

Obtaining those measurements can be tricky. While some birds are relatively calm, others are spirited. American kestrels, for example, are infamous for pecking. Approximately the size of a robin, these feisty falcons nip at our fingers as we extricate them from the net, focusing in particular on any hangnails they can reach. (We don't wear gloves because we need to be able to feel as well as see what we're doing, and gloves would dull our sense of touch.)

Kestrels peck at the calipers as we adjust the moveable end and try to align it with the exact tip of the bird's upper mandible so we can measure the length of the beak. They bite the little plastic ruler we slip beneath their wings to measure the distance to the tip of the longest wing feather. And who can blame them?

This poking and prodding is akin to what humans might undergo at an annual physical exam in the doctor's office. While no one would characterize it as enjoyable, it's usually only mildly annoying, and as with humans, it's important for monitoring the health of both the individual and the species. Carnivorous raptors, being high up on the food chain, warn us of contaminants like DDT—and, more recently, rat poison—that have worked their way into the environment and threaten many species.

It takes hundreds of volunteer field biologists to gather all this avian data. We banders work every single day of the week during the August-to-December migration season, catching hawks during most of the daylight hours. In just over thirty years, the GGRO has trapped nearly 40,000 hawks, keeping meticulous records about each one. But the bird I was concerned about that day was the one gripping my thigh.

"Buzz, when you have a moment, I could use some help." I spoke softly and slowly. Buzz is not the kind of person who makes sudden movements, but I wasn't taking any chances. A sudden movement would cause certain pain.

"OK; give me a minute here." Buzz was bent over a Cooper's hawk he had just caught, measuring its wing length. He did not look up.

I decided to wait for Buzz to finish his work before explaining my predicament. He worked quickly; it would only be a few minutes. The sound of our voices made my bird nervous, and when she felt nervous she tightened her grip. As long as the bird and I were both relaxed, and no one was moving or speaking, the pain was bearable. Once Buzz finished with his "Coop," we could focus on my red-tail.

I felt a little light-headed and needed to distract myself from the pain, so I looked more closely at the raptor in my lap. Her breast feathers were bright white; the stomach a bit darker, with buff-colored streaks. Her back was chocolate-brown, and the tail dark brown with paler thin bands. (This was a first-year bird; the distinctive brick-red tail feathers would not grow in until the following year.) All the feathers were in good condition, and I saw no lice or other ectoparasites; that was a good sign.

I didn't want to look directly into the hawk's eyes, thinking that might antagonize her. But I couldn't help myself. They were mesmerizing: fierce, bright and alert, with pale golden irises.

When I saw that Buzz was nearly finished with his Cooper's hawk, I continued. "I've been footed." *Footed* is the technical term for having a hawk's talon deeply embedded in some part of your body. It's embarrassing, because it means you haven't been paying close enough attention to what you're doing. You have not properly controlled the hawk's feet. You have made a stupid mistake. The only good thing about being footed is that it hurts so much you will probably never let it happen again.

Buzz turned toward me slowly, assessing the situation. It may have taken a second for him to move to my side, or maybe it took a year. It was certainly a good ten minutes before he was able to pry open the hawk's clenched foot and slide the curved talon out of my leg.

Liberated from the red-tail's grasp at last, I lay down on the rough wooden floor of the blind and closed my eyes. Buzz stepped outside and released the red-tail, which I'm sure seemed unconcerned as she flew off. They always are. Often a hawk will even give a quick little shiver as it flaps away, as if to shake off the memory of a close encounter with humans.

The pain subsided quickly, and soon a deep puncture wound was all that remained of my avian adventure. That was ten years ago. But even though I was freed from that hawk's grasp, I will never be free from my fascination with these wild creatures. Something about them captured my imagination and pierced me to the quick. And every time I look into a hawk's eyes, I am hooked all over again.

:::::

In the pages ahead, you'll meet more animals that will make you gasp, laugh, and wonder—some so close you can feel their breath, others so strange you'll hardly believe they exist. Each story stands alone, yet together they reveal the wild world in ways you won't forget.

A big cat that does not look particularly hungry—
but who knows for sure?

Big Cats, No Guns

On foot with lions in the Okavango Delta

The first time I tracked lions, I sat safely and comfortably in a large, open Land Rover, with a loaded rifle next to the driver. At that time, our guide had assured us that as long as we didn't wear brightly colored clothes, make noise, or stand up, the animals would perceive us as part of the vehicle, and therefore not worth eating. His logic was not convincing. Lions have been making their living—for two million years—by figuring out what is, or is not, edible. And we were going to fool them by sitting instead of standing? I was sure the big cats were smarter than that.

But this safari was different. We were going on foot, and the strict policy at Camp Okavango was *no guns*. Big cats, no guns, traveling on foot . . . hmmm.

Adding to my trepidation, our guides, Robert and Rodgers, explained that if we saw lions this morning, they would be hungry, because big cats usually hunt at night. If they were still out stalking prey in the morning, it meant they had found nothing to eat the night before. A crazy thought edged its way into my mind: The guides were using us as lion bait. They thought we were clueless American tourists, foolish enough to follow them deep into the big cats' territory with no means of protection.

We were.

Their plan was to travel from our camp by motorboat through the vast delta to another island, and from there to proceed on foot in search of the big cats. Many miles out, through the winding, papyrus-lined waterways, Robert announced excitedly that he had spotted dust in the trees. I didn't see it, even with my binoculars. And I didn't understand what dust had to do with lions. But I went along with the program. We anchored the boat, disembarked, and walked into the remote island's open forest.

This was no Sunday stroll: The tall brown grasses hid treacherous obstacles. Elephants had eaten the relatively tender bark and roots of trees, leaving dead branches and uprooted stumps scattered everywhere. Aardvarks had dug large holes in the ground. Thorns caught on our clothing, and greedy vines grabbed at our legs. And the dung! Everywhere, we had to step over dung, all kinds of it: large and small, round and elongated, fresh and dry, in varying stages of decomposition.

I could tell the difference between rhino middens and elephant dung, and was learning to differentiate buffalo from giraffe. Then I saw a new kind: smaller, rounder, fresher—glistening, in fact. Was it lions'? I wondered just how far away the lions actually were, and how close we intended to get. Quickly checking my field guide, I found that lion droppings "are similar to that of the leopard, but larger." This was only marginally more helpful than the entry for elephants, which read, "A good way of testing the freshness of dung is to thrust your hand into the centre of it. If the dung is fresh, it will be warm inside." Right. Like I'm ever going to employ that methodology. (They didn't even provide a chart correlating temperature to time elapsed to distance traveled.)

We were a noisy bunch of Americans, and Rodgers admonished us to "do *shhh*" and to "talk silently." I noticed that Rodgers and Robert did indeed talk silently, communicating with their eyes and hands that they had heard a noise in this direction, or that they wanted us to walk that way. They reminded me of a TV SWAT team, moving swiftly and efficiently through the bad guys' hideout just before the big gunfight. Our group, on the other hand, moved like a bunch of Keystone Cops, zigzagging randomly, tripping in the aardvark holes, backtracking around fallen trees, and fighting back with all our might when vicious vines attacked.

Robert reminded us to walk in single file, always staying together. If we fell back or got out of line, he warned, we'd look smaller and be "on the menu." *As if lions haven't had plenty of experience in picking individuals out of a herd,* I thought. *As if they would look hungrily at a line of humans hiking single file and decide, "That's just some giant caterpillar, over-burdened with indigestible cameras and binoculars, stumbling slowly and vulnerably through my sovereign territory. Forget it—no chow there."*

Most important, Robert said: if a lion did come toward us, "Don't run! Stand your ground!"

Stand my ground? In the face of a charging lion? What kind of instruction was that? My heart pounded at the thought of it; my legs stiffened, and I wondered whether it would be a good thing to be frozen with fear. My hands began to sweat, and I remembered reading somewhere that humans, dogs, and other mammals whose paws sweat with anxiety do so because the sweat increases friction between the paws and a substrate, allowing for quicker getaways. I was built to run, then, *not* to stand my ground!

I stayed near the front of the line, just behind Maureen, a friendly psychiatric nurse from Pittsburgh. Maureen walked with her shoulders slightly hunched, took each step slowly and deliberately, and shaded her eyes from the sun as she scanned the distance.

I thought she looked like a professional tracker, except for her bright white Asics Gel running shoes—obviously bought new for the trip—and hot pink windbreaker. I had chosen my place carefully: Surely the hot-pink-windbreaker variety of meal would be most tempting. If a lion charged, I would simply maintain my position behind the primary bait.

Maureen's left shoelace was untied. Should I tell her? If I did, she'd stop to tie it, and the whole single-file line would crash into us like a row of dominoes. I would be trampled by my fellow travelers, and perhaps sprain my ankle or fall into a pile of warm dung in the process. If I didn't, Maureen might trip and fall, and be eaten alive. I kept my mouth shut.

A large, lone bird circled the sky above us. Robert identified it as a white-backed vulture *(Gyps africanus)*. Our guides had amazing eyesight. Born and raised in the delta, they could identify its birds and animals from far away. Soon a second vulture appeared, and then a third. Apparently they knew something we did not.

As we hiked deeper into the forest—way too far in to run for the boat—Robert still saw dust. We kept walking, single file, assiduously staying together, not falling back or getting out of line, doing *shhh*, going deeper into the forest.

At one point I was tempted to stop and photograph

a little bee-eater (*Merops pusillus*), a tiny, brilliantly colored and exquisitely beautiful bird, as it caught insects in the tall grass. But when I considered the possibility of "death by lion," I decided to stay with the group.

"Listen!" Rodgers and Robert both heard the lions. We kept walking. At first we didn't hear anything, but farther in we heard a low rumbling sound. "It's the lions! Yes, and they are chasing buffalo!" The rumbling, our guides explained excitedly, was the sound of a thousand hooves. We proceeded, still in line, straining to get a look through the trees at a buffalo or a lion. Suddenly Robert hurried back into our midst, eyes wide and bulging so the whites showed around their whole circumference. "They are coming this way!" he shouted hoarsely. "*We are too close! Go back! Go back!*"

Finally I saw the dust, a huge cloud of it, about 200 yards away and coming toward us fast. It was swirling above a herd of several hundred Cape buffalo, and they were coming toward us fast, too. Stampeding, actually.

We had been told to stand our ground in the face of a charging lion, but what was the protocol for a buffalo attack? There was no time to ask. Our careful single-file line disintegrated into chaos as we ran back—hats, cameras, and binoculars flying. No more zigzagging to avoid holes or backtracking around fallen trees; we leapt them all heroically. Several of our group turned out to be talented sprinters, and I personally tested the freshness of five or six piles of dung in the space of twenty seconds.

As suddenly as the stampede began, it was over. It's interesting what goes through one's mind at a time like this. As soon as *"Escape, escape!"* had run its course, I was overwhelmed with the perfection of Nature's Grand Plan: Elephants knock down large trees, allowing grasslands to develop, which attracts grazing animals, which provide food for the lions. The aardvark holes create natural traps for the lions' prey; the elephants' monumental, nutrient-rich droppings fertilize the tall grasses.... Lost in the beauty of the Grand Plan, it was several minutes before I remembered Maureen. How did her untied shoelace fit in? Had I been homicidally remiss in not mentioning it earlier? Or was I simply playing my predetermined part in the "survival of the fittest"? Did Maureen stand, or did she run? Had she been trampled by stampeding buffalo, or eaten by a hungry lion?

I came to my senses, surveyed the scene, and saw Maureen's hot pink jacket halfway up a small tree, with Maureen still inside it. Apparently it had not provoked the lions. We began to regroup, and everyone seemed to have survived. The Cape buffalo—still about 70 yards away—had also survived. They had all stopped running and were now milling about restlessly. They seemed to be more afraid of *us* than of the charging lions. This did not strike me as an effective adaptive behavior, but what do I know about the life of a buffalo? And what did they know about humans? At any rate, they kept their eyes on us *and* on the lions, which—conveniently—made it easy for us to observe the five adult lions that were now in our immediate vicinity.

Make that five *hungry* adult lions.

I had heard that female lions form hunting bands, and that the males don't bother to assist them. But this was a group of four females and one large male with an impressive golden mane. What was *he* doing there? He was probably so famished that he couldn't even wait for the females to make a kill. He was big, that was for sure. I couldn't see his teeth, but I knew they were long and sharp, and I'll bet he was salivating.

I smelled the dust. Dust, and a sweat-like smell. Was it the buffalo, or the lions? Was it me? Was it fear? The lions were not yet attacking, so I had a moment to contemplate: Should I turn and run, or stand and scream? Would it make any difference? Did I have a choice? I would like to report that Rodgers and Robert were unfazed, but that would not be entirely true. Actually, they looked anxious. They had no guns.

They were responsible for a dozen travelers. And we were 70 yards away from five hungry lions.

The big cats paced around the edges of the buffalo herd, eyeing one individual, then another. What if one looked at me? Should I make eye contact, or avoid it? I began to feel panicky. Robert's words echoed in my mind: "If a lion does come toward you, *don't run!*" Legs locked, I stood my ground, and felt proud of myself for having the presence of mind to follow instructions in a crisis. But then it occurred to me: If I stood my ground, and everyone else retreated, did that make me clever? Or did it make me lion food?

Fortunately, my survival did not depend on my own cleverness: Rodgers and Robert took control, instructing us to remain facing the lions and walk slowly backwards, away from them. We did so. Soon, it was safe to turn and walk—a bit more quickly—back to the boat. When we were safely on board, and motoring back to camp, I marveled at the beauty of Nature's Grand Plan, at green papyrus against intense blue skies and the exquisite deliciousness of a bottle of cold beer. Life is good, when you're not on the menu.

Quest for the Karkadann

Tracking Namibia's rare black rhino

The *karkadann* was once shrouded in myth. Ancient texts describe a creature swift and powerful, fierce and mystical—and impossible to capture alive. Its long horn was said to purify water and neutralize poisons. As long ago as 400 BCE, the Greek historian Ctesias wrote of a white-bodied beast with a cubit-long horn, red at the tip, black in the middle, and white at the base, with eyes as blue as sapphires:

> "Those who drink from these horns,
> made into drinking vessels, are ...
> immune even to poisons if, either
> before or after swallowing such, they
> drink wine, water, or anything else
> from these beakers."

Medieval unicorns undoubtedly descended from this legend and, as well-established Christian religious symbols, were identified with chivalry, purity, and chastity. It was said that unicorns could only be subdued by a virgin—a reference to the Virgin Mary.

It's hard to imagine how the rhinoceros, hefty of body and somewhat stubby of horn, came to be conflated with the lithe unicorn.

In modern Iraq, too, "tears of the karkadann" have religious symbology. They are reddish prayer beads, said to have been created when the tears of the beast, weeping "out of fatigue and thirst-pain," fell into the water of a drinking hole.

Karkadann is the Persian word for rhinoceros, and rhino horns have enchanted us humans for at least as long as unicorns have. Chinese texts say they allow the bearer to breathe underwater. Persians of the fifth century BCE believed vessels carved from the horn would cause liquids to bubble if poison was present, which would certainly have made them a sought-after commodity.

The belief persisted in European courts as late as the nineteenth century. I can imagine a fearful royal, wary of palace intrigue, calling for his karkadann horn before a night of revelry. It is also revered in many societies for its purported aphrodisiacal qualities.

In fact, there may be an element of truth behind the rhino horn's supposed ability to detect poisons. They are composed largely of keratin, the same protein that makes up our hair and fingernails—not to mention horses' hooves, turtle beaks, and cockatoo bills. Many poisons are strongly alkaline and could have reacted chemically with the keratin.

But myths aside, the true karkadann lives on, Critically Endangered, in Africa's grasslands and deserts. It is a creature of flesh, not fantasy—massive, armored, and unpredictable. I traveled to Namibia to see it. My local guide, Me-Gusto, arranged for us to tag along with two professional rhino trackers who work for Save the Rhino, a conservation organization. He asked that I not report details about how many rhinos we saw in which areas of Namibia, because poaching is a problem: Locals, confronted with limited economic opportunities, have often turned to plunder for food and income. It is best not to advertise the rhinos' whereabouts.

I *can* say that we visited Etosha National Park in northwestern Namibia. The park's diverse vegetation ranges from dense bush to open plains, and in its heart the famous Etosha Pan—"the place of dry water"—is a shallow depression covering nearly 2,000 square miles. Dry and shimmering with salt deposits for most of the year, its perennial springs and waterholes attract lots of wildlife.

We also drove through the 115-square-mile private Ongava Game Reserve, which butts up to Etosha National Park's southern boundary and has a similar ecology. Ongava (the name means "rhinoceros" in the local Otjiherero language) is notable for being home to the largest population of southwestern black rhinos in Africa. A third area we checked out is Palmwag, a nature reserve located farther west in Damaraland, which is an even harsher environment. We saw rhinos in all three locations, along with elephants, kudu, giraffe, hartebeest, springboks, impala, oryx, zebras, gnu, jackals, and hyenas in various combinations.

Most of the rhinos we saw were black rhinos, which were once found throughout sub-Saharan Africa. European hunters and settlers nearly killed them off in the twentieth century, but persistent conservation efforts across Africa have helped their numbers rebound.

The World Economic Forum reports that there are at most 5,458 black rhinos and 21,085 white rhinos left in the wild. Me-Gusto doubts that those numbers are accurate. As a long-time guide, he has known plenty of rhino trackers. "Where do you think the conservation organizations get their information?" he asks. "A success story would bring too much unwanted attention to the area. But if trackers report numbers that are too low, NGOs conclude that their efforts are not working, and divert funds to areas where they believe they can make more of a difference."

No matter what the exact numbers are, the black rhino is still Critically Endangered, and a lot of work remains to bring the population up to even a fraction of what it once was—and to ensure that it stays there. Wildlife crime continues to plague the species and threaten its recovery.

By the way, that's *southern* white rhinos. The genetically distinct northern white rhino subspecies is functionally extinct; only two individuals, a mother and daughter, remain. They live on a conservancy in Kenya and are protected round-the-clock by armed guards.

Understanding the challenges they face made me even more eager to see the rhinos. The first excursion was easy. It was a June day, winter in Namibia, and the temperatures were in the pleasant high 60s when we set out at six a.m., hoping to find a rhino that was still out in the cool morning air. Namibian rhinos are desert-adapted, able to survive the region's extreme temperatures, ranging from below zero to more than 104°F. To beat the brutal daytime heat, they are most active at night, in the early morning, and in late afternoon.

After bumping along a severely rutted path for many miles and then turning off-road, our rhino-tracking guides Joel and Michael located a female called Missy. She had been dehorned and was traveling with her calf. We watched the two rhinos for half an hour as they stood quietly, ears and nostrils alert, browsing on the sparse shrubbery.

They didn't do much, and I had to remind myself that they were not obligated to "do" anything for me. Nevertheless, observing such rare creatures in the wild was mesmerizing. The massive bulk of the rhino's body and desert-hardened skin were tangible evidence of evolution at work. Michael showed me the booklet of ear-notch images they use to identify individual rhinos, and the forms they complete to record the time and GPS coordinates of the sighting, as well as information about the rhino's general condition.

Me-Gusto explained the dehorning. "Since poaching for rhino horns is a significant threat," he said, "conservationists tried dehorning the rhinos to save their lives. It seemed like a good idea," he continued, "but it didn't work. In fact, the plan backfired."

"But I've read that dehorning was effective," I protested.

"You've read a lot of things," Me-Gusto said. "Do you believe them all? Sometimes a decrease in poaching in one place just means the hunters have moved on to a new location. Rhino poaching is hard, hot work. It involves walking for many miles a day, far from civilization, and camping at night with only the provisions you have carried on your back through the desert.

This might go on for as long as twenty days at a time because, especially with recent droughts, the rhinos have to travel through massive areas to find enough food; a rhino's home range can be larger than 200 square miles. Their main source of food, the highly toxic desert plant *Euphorbia damarana*, is deadly for humans and other animals." Me-Gusto knew the desert plants as well as the animals.

"Poaching also requires the ability to read spoor and tracks, and a talent for befriending local people who might have news of the wildlife's latest movements in the area," Me-Gusto continued. "And it requires evading the local anti-poaching patrols, like Joel and Michael.

"So a poacher has a lot invested in each hunt. If he tracks a rhino for days or weeks, only to discover that it has been dehorned, he's going to be disappointed.

"He goes off in search of another rhino, only to find that it, too, has been de-horned. He may wonder: Is this the same rhino as the first one, or a new dehorned rhino? He'll likely be angry. And hungry. The rhino horn would have been his payday; he has worked for weeks and has nothing to show for it." Rhino horns can fetch upwards of $27,000 per pound on the black market—far more than elephant ivory.

"At this point he may decide it makes sense to shoot any dehorned rhinos he comes across, so he doesn't waste valuable time tracking them in the future." The dehorning project, promising as it sounded, turned out to be a spectacular misfire.

Fortunately, there are alternatives. In addition to education and anti-poaching squads, conservationists are experimenting with radio-tracking collars, AI-enabled drones, radioactive isotopes inserted into horns, 3D-printed synthetic horns, bio-engineered lab-grown horns, and several other methodologies. I even read that a Swedish wildlife photographer has been collecting donated fingernails to replicate a popular Asian medicine made from rhino horns. "We're going to [sell it to] China and Vietnam as a substitute," he says.

A few days later I got the thrill of seeing a rhino up close—maybe a little too close—on foot. We hadn't yet had a good sighting at this particular location, and James, the naturalist who was taking us out that day, was determined to provide the experience we had come for. He picked us up in an open-top Land Cruiser, perfect for wildlife viewing. And he brought along a .375 rifle.

I looked it up, and a .375 is standard for "last resort defense against dangerous class 3" animals such as "thin-skinned game averaging 300-2,000 pounds live weight." This did not seem even remotely adequate to me, since an adult male black rhino can weigh upwards of 3,000 pounds. White rhinos are even bigger: The males average 5,100 pounds. Also, their skin is anything but thin. Especially these desert rhinos, which have evolved with extra-tough hides to protect them from the desert climate.

Aside from what I looked up, I know absolutely nothing about guns, or shooting, or calibers, or ammunition, or armor-plated behemoths. I began to wonder how safe our adventure would be. Adding to my discomfort, my body was changing in the arid climate: My eyes were constantly squinted, to keep out both the blowing sand and the bright sunlight. They reminded me of the tiny eyes of the rhinos and elephants we saw. The skin on my thighs had begun to wrinkle in the dry air, making them look increasingly like rhino hide. And my fingernails became so brittle they broke off every day, so I could literally count my vacation days on my fingers.

We bounced along for hours through the desert, jarring our bones and breathing in buckets of alkaline dust, stopping occasionally while James got out to search for tracks in promising locations. But we found no rhinos. *I needn't have worried about danger,* I thought, disappointed with our afternoon safari.

I entertained myself by recalling information I'd discovered about uses of rhino horns. Even today, Traditional Chinese Medicine recommends small amounts, ground into a powder and dissolved in boiling water, for treatment of fever, rheumatism, and gout. This may have some basis in fact: Toward the end of the twentieth century, researchers in Hong Kong found that very large doses of rhino horn could slightly lower a fever in rats. The necessary dosage was so high, however, that "you'd do just as well chewing on your fingernails."

The rhino horn has long been used medicinally as well as magically. The 1597 *materia medica* written by Chinese pharmacist Li Shih-Chen alleges that it could "cure snakebites, typhoid, hallucinations, headaches, carbuncles, vomiting, food poisoning, and devil possession."

Nevertheless, many Asian countries, from India and Malaysia to China and South Korea, use various parts of the rhino to treat a range of ailments. Even the blood, urine, and dung are used. Employees at Chitwan National Park and the Kathmandu Zoo in Nepal, for example, are said to surreptitiously collect rhinoceros urine, which some people drink to relieve asthma attacks.

I'm curious to know how that would work. Whether in a zoo or out on the range, getting right up close to a rhino, even from behind, is surely a perilous proposition. Not to mention the process of collection: knowing when the rhino was about to relieve itself, maneuvering the container into the appropriate position, adjusting for the trajectory and velocity of the stream, holding the increasingly heavy vessel at arm's length without spilling the precious fluid, and then withdrawing it when the container was full and making a safe and speedy getaway.

Selling any rhino parts—or products—is highly illegal in Nepal, so there would be the usual black market complications to deal with, as well. Add to this the fact that alpha male Indian rhinos can spray their urine more than sixteen feet, and one can only imagine the risk and discomfort involved. It would take time, too, since a rhino can urinate up to five gallons in a single session. However, my calculations suggest that the proceeds of just one tinkle from a single large rhino per day could more than double a guard's salary, so one can see how the economics, if nothing else, might be appealing.

It was nearly sundown, the wide sky a magnificent palette of pink and purple. I had given up hope of spotting a rhino that day. James wanted to look in just one more place, however, and followed a hunch through several miles of low sand dunes and sparse bushes, finally lurching across a series of boulders in a dried-up streambed and jolting to a stop in a shallow gully. Sure enough, he had found a solitary rhino, some 400 yards in the distance—too far for a good look, even with my 15x binoculars. He inched the Land Cruiser closer, but the rhino meandered away.

"Rhinos are shy animals," Me-Gusto noted. "They're easily spooked."

James drove a few yards closer, and the pattern repeated. The rhino didn't seem upset, but neither did it let the vehicle get close. We tried a third time, with the same result. Finally James explained, "The only way to see rhinos up close is to walk to where they are. Are you up for it?"

We hopped out of the Land Cruiser. Down on the ground, our view wasn't as good. In fact, we could no longer even see the rhino. But we walked in the direction we had last seen it, weaving around boulders and peering through the brush, hoping for a closer look. Fine, dry sand shifted beneath our feet, slowing our progress. It would have made a rapid getaway, should we need to execute one, all but impossible. A high salt content had turned the short tufts of grass so brittle they crunched loudly underfoot, announcing our approach. It sounded as though we were walking on a crust of snow.

We trudged the length of a football field without seeing or hearing the rhino. I suspected it had wandered away. Either that, or it would suddenly come charging out of the nearby brush to gore us with its magical horn, the way unicorns do in movies. Survival strategies began to run through my mind as I chewed nervously on my fingernails. Not being a virgin, I would be unable to subdue the beast myself. I stayed close to the others.

"Rhinos are often thought to be solitary," Me-Gusto whispered, continuing the never-ending wildlife commentary he knew I loved. "But actually they are loosely social; they even tolerate the presence of unrelated rhinos at a waterhole. So if you see what seems to be a solitary rhino, it is almost invariably part of a group, and there are likely more to be seen nearby." It was like having my own personal Nature Channel narrator.

The sky turned purple as we rounded a bend, and there they were—not one but two huge rhinos, glowing lavender in the fading evening, eyeing us warily. Less than thirty yards away, they flinched but held their ground. We did the same. "We are close enough," James said. "Rhinos don't see well. But we're upwind. They know we are here."

We stood there with the rhinos for twenty minutes or so, watching each other as long as the light lasted, neither advancing nor retreating. The experience did indeed seem magical, until the larger rhino turned his gaze directly at us and took a step in our direction. It was not a tentative step. His front left foot crushed the grass purposefully. But he didn't stop there. His ton-and-a-half body barreled forward.

"Stand still!" James commanded. "He cannot see us if we stand still!" James did not seem worried. He did not raise his .375 rifle. No one moved. Not a word, not a flinch. The hulking creature moved surprisingly quickly, but after a few more steps it stopped, looking confused. Well, as confused as an inscrutable megamammal can look. Apparently, he had lost sight of us in the dimming light. The rhino turned slowly and ambled away. His companion had already moved off.

As they disappeared into the twilight, the magnitude of the encounter struck me. These are animals that inspire myth, awe, and reverence, yet their existence is precarious. Seeing them was not just an adventure—it was a reminder of why conservation matters.

Despite his comical appearance, this cassowary can kill a human.
With his toenails.

The Truth About Eco-Travel

At one with the rainforest in Cape Tribulation, Australia

Jim was hesitant from the start. "Cape Tribulation? *Wilderness Area?* No way!"

Our travel agent had provided a glossy brochure of the Bunyip Lodge, and I cajoled my husband into going along to this eco-resort in northeast Australia. "Eco-tourism" sounded so romantic: waking to the trill of morning birdsong, viewing exotic animals without binoculars, and falling asleep to the melodic sounds of the night forest. Who could resist two weeks in a pristine rainforest? I reminded Jim that there was a *resort* in eco-resort, and assured him that any inconveniences would be minor. Besides, the brochure featured a photo of a sparkling swimming pool. If Jim decided against traipsing through the rainforest with me, he could always relax by the water.

I wanted to be one with the rainforest.

We traveled twelve thousand miles by air and bus, transferred to a treacherous wooden ferry to cross the Daintree River's gaping mouth, and finished our journey in an eight-passenger minibus that rattled and shivered and shimmied along a deeply rutted road. So far, so good. But, arriving at the Bunyip Lodge in the late evening, we began to discover the truth about eco-tourism. The minibus would not fit on the "resort's" eco-friendly-but-overgrown road, so the driver was forced to deposit us—rather unceremoniously, I thought—at the side of the lane that cut a thin ribbon through dense jungle. There were no lights at the unmarked drop-off point and, as the minibus quivered off to its next destination, we were left alone—in complete darkness—with our luggage.

We hadn't packed flashlights. Who would think we'd need one to get to the hotel lobby? Jim and I stood stupidly in place for a few minutes as our eyes adjusted, and eventually discovered a path leading through the jungle to a faint light. Leaving our luggage heaped in a pile, we stumbled down the bumpy, vine-tangled trail toward the resort. *The porters can retrieve it later*, I thought.

Although there was no lobby, we did manage to track down the proprietor, an amiable fellow called Tony, who was unable to check us in because the computer was down.

"A bug?" I sympathized, remembering the last time my own computer had crashed.

"Nawr—it's the bloody mice. They've chewed right through the wires again," Tony explained. Ever helpful, he lent us a couple of torches so we could retrieve our luggage and carry it to the cabin.

There were no porters.

Jim and I were both dripping with sweat before we had walked the hundred yards from the drop-off point to the office. But we remained optimistic and decided that although the weather and accommodations were not exactly what we had anticipated, it would be fine as long as there was air conditioning.

Of course, there was none.

There *were* cracks in the walls wide enough for a small dog to pass through—surely these were not for wildlife viewing—and a narrow bed surrounded by alarming volumes of mosquito netting (never a good sign). There was also a plasticized placard in the room requesting that we leave the lights and ceiling fan on at all times to inhibit the encroachment of creeping jungle rot. Occupying a prominent position on the tiny bathroom counter was a super-sized red box of salt, along with instructions for using it to remove leeches: "Simply rub the salt over the attached leech...."

Jim and I sank onto the bed and turned to face each other. He beat me to the question: "Whose idea *was* this, anyway?"

It was time for some serious attitude adjustment. We ambled over to the resort's open-air bar, drank cold gin and tonics, and leafed through the limp pages of several field guides we found stacked on the barstools —a makeshift library. (Suddenly, I understood why the local currency had a high plastic content: Plain paper does not hold up in such high humidity.) Bats swooped precipitously overhead.

I planned to search for the rare buff-breasted paradise kingfisher. "It migrates all the way from New Guinea," the guidebook said, "to breed only in this small area in North Queensland, and nests by burrowing into termite mounds." I would also be on the lookout for the musky rat kangaroo, a "rat-sized" marsupial that looks like a miniature kangaroo, climbs trees, hops like a rabbit, and commonly nests in the dreaded "wait-a-while" plant. The third animal I hoped to spot was the cassowary, an ostrich-like bird that can kick a person to death in self-defense. It wasn't entirely clear, from the text, whether death resulted from the power of the kick itself or from the sharp nails on the bird's inner toes, which "can easily rip human flesh." Now *this* was exciting.

The cassowary, I learned, is Australia's largest land animal, weighing as much as 130 pounds. Not surprisingly, it is quite incapable of flight. Vestigial wings each carry three to five wire-like feathers, which are used to help brush aside undergrowth as the cassowary travels through the rainforest. As it moves, the giant bird lowers its head "for protection" and "lifts its toes right up under its chin." I couldn't fathom what a gigantic, man-killing bird might need protection from—except, perhaps, humans, as it is an endangered species, with fewer than 1,500 individuals remaining. The male cassowary incubates eggs and cares for the young, which I hope bodes well for the survival of the species.

But larger-than-human size, a comical gait, and killer toenails are just the beginning of the cassowary's odd characteristics. It also sports body feathers that look like foot-and-a-half-long strands of luxurious taupe-colored fur, a bright blue neck, a long red wattle, and a head crowned with a large casque or "helmet" of horn-covered cartilage, the size of which is believed to be significant in determining the cassowary's social status.

And let's not forget the bird's "uniquely short digestive system," which allows it to eat the fruits of poisonous plants and eliminate the toxins before absorbing them. In fact, seeds often remain intact, and can grow after passing through the cassowary, which plays an essential part in the dispersion of seeds from up to 100 species of trees and shrubs. Other rainforest inhabitants—notably the musky rat kangaroo—

commonly include partially digested fruit from the cassowary's droppings in their diets. This bird—shy, fast-moving, and integral to the ecosystem—had captured my imagination. Here was a creature worth viewing, and I was eager to begin the search.

Not wanting to be outdone, Jim studied up on the local flora. "It says there's something called the gimpy-gimpy plant that causes horses to commit suicide."

"Does not," I countered. Surely he was kidding.

"*Duh-zz*. Page 137. The most terrifying plant in the area," Jim read, "is the gimpy-gimpy. When brushed against, its leaves release an extremely painful irritant. There is no known antidote, and the pain can last for months."

"All right, but what about the horses?" I interrupted, catching him in what was surely a complete fabrication.

"Horses have been known to die from charging into trees after exposure to this plant," he continued. "The leaves appear soft and fuzzy from a distance, and have been used, by some stunningly unfortunate explorers, as rainforest toilet paper."

Ouch.

Second on the dangerous-to-horses-and-other-large-mammals list is the "wait-a-while" tree, also called "lawyer cane" because, once hooked by the thorns on this pitiless plant, one is as irretrievably entangled as if involved in the legal process. The vegetation starts out looking like a small palm tree, then grows long, wiry "tendrils" which are decidedly not tender—they're lined with rows of viciously sharp barbed hooks. If you happen to brush past one of these possessive forest dwellers, it grabs your clothing and holds on tightly until you back up and remove the thorns. The tree has even been known to "pull people off horses and motorcycles" as they ride by. Eventually, the long, barbed, vine-like part grows to an inch or more in diameter, the thorns drop off, and the result is the smoothly benign rattan from which patio chairs were made in the 1970s.

Over another round of gin and tonics, Jim and I anticipated the next day's activities, which now revolved around avoiding terrifying plant life. "We might be lucky enough to spot a dusky-colored, rat-like kangaroo scampering about in the rainforest shadows," I enthused. "It's the world's most primitive marsupial!"

"Yes, and we could also see a small brown bird covered with termites, or a large brown bird that could kick us to death," Jim replied, with somewhat less exuberance. "*These* were the exotic native species you had dreamed of encountering?"

Next morning, the air remained thick. The shirts we had worn the day before and hung in front of the window to air out remained wet—not merely damp, but soaked through—with perspiration.

Everything inside our suitcases was wet, too, so Jim used a hairdryer on our shorts and shirts. Then we headed for the lodge, where I scouted around for the inviting swimming pool that had been featured in the brochure. Desperate to float in it, I imagined diving in, a brisk splash, then full-body relief.

The pool did not exist.

Oh, there was a tiny wading hole, similar in overall look, if not dimension, to the palatial pool featured in the brochure. The photographer had apparently taken the shot from ground level and used an extremely wide-angle lens in order to flatter the tiny pond and lure unsuspecting tourists.

Tony was sympathetic and let us in on a secret, suggesting that we spend the morning "cooling your inner core," as the locals did, by soaking in Cooper's Creek just down the hill. He assured us that the leeches there were only small, the kind that bite you gently and then fall off. "Just a little nibble, really. No worries."

The photo-perfect swimming hole was clear and beautiful, surrounded by lush tropical vegetation, shaded by magnificent rainforest trees. And the water was cold, as Tony had promised.

Very, very cold.

I'm a bit of a baby about plunging my body into

frigid, leech-filled streams, so it took me a good twenty minutes to submerge. My feet were easy, lower legs not so bad. Thighs difficult, waist nearly impossible. The water was unbearably cold, and I kept thinking about the leeches, and which of my body parts they would be most likely to attach themselves to. Did they prefer light or dark? Cold or warmth? Freedom to crawl around or the protection of a cozy nook or cranny?

With my mind finally off the oppressive heat I relaxed, looked around, and began to think: *Who would live in a place like this, anyway? Were they all insane? Had they been kicked out of other towns, or even other countries?* Visions of early shiploads of convicts filled my mind. Perhaps the crazy ones had migrated to the hinterlands of Queensland? But no—everyone we had met seemed quite civilized.

An intense vibration interrupted my reverie, and I realized my teeth were chattering uncontrollably. After sitting in the numbingly cold creek for more than an hour, we had unwittingly induced hypothermia. *This* was the locals' strategy for surviving in a sweltering climate!

But we did finally feel better—especially after removing the small leech that clung tenderly to my right big toe. So much better, in fact, that we decided to take a nature walk, hoping to spot an elusive cassowary. Tony drove us to a nearby reserve and introduced us to Helen, a stout, Hobbit-like woman with the air of someone who had spent a few too many years alone in the jungle. Although the sun was shining and the blue sky showed no hint of rain, she dressed in a bright yellow slicker and knee-high gumboots to guide our rainforest walk.

We traipsed for hours through lush jungle suffused with dim green luminescence. In some places the sun barely filtered through, and light seemed to emanate from mossy surfaces and dew-covered leaves. Primeval cycads—primitive plants whose existence dates back 250 million years—stood in groves, water dripping from their palm-like fronds. Ulysses butterflies, brilliant blue, with four-inch wingspans, floated silently by. I was in heaven.

Jim did not share my appreciation of the living museum surrounding us, home to some of the world's rarest plants and animals. "I think we've seen all 3,000 endemic species," he muttered from behind me on the trail. "Now I only need to sit on a gimpy-gimpy plant to make my trip complete."

Then we came upon what were clearly Helen's favorites: ants swarming by the thousands along rough tree branches. Each ant was about a quarter of an inch long, and had a delicate, bright green abdomen that was grossly overextended—filled with nectar the ant had collected—and looked as though it was about to burst. Helen gently pulled an extra-large ant off a tree and extended it, posterior end first, to within inches of my face. "'Ere ya go, then. Go ahead, lick its butt."

Lick the ant's butt?

I don't know whether I was more afraid of the ant, with its horrifyingly engorged abdomen, or of Helen and her curiously insistent attitude. She touched the ant's translucent bulge to her own confidently outstretched tongue, and I bravely followed suit. After it was clear I had not been poisoned, Jim licked his own ant, and his eyes shot open. The taste was like mixing the intense fizz of an Alka-Seltzer with tangy lime sherbet.

Helen explained that the ants were great to have around because they kept other bugs away. I was glad to hear this, as I'm always the first to know when mosquitoes are nearby: They apparently sense the sweetness of my blood, or the depth of my hatred for their species, and swarm about me without fail.

Despite the ants, today was no exception. The mosquitoes swarmed, landed, and sucked my blood mercilessly. We convinced Helen to turn back just as it began to rain. Venomous tree frogs croaked a deafening chorus that echoed in every direction and disoriented me completely. Water gushed down in torrents, and my respect for Helen's sartorial eccentricities increased. Our feet sank into calf-deep mud, which threatened my balance and oozed its way down to the very toes of my new Blundstone hiking boots. My bare arms and legs itched wildly.

And that's when I realized my wish had been granted.

Although we had not spotted the elusive cassowary, I *had* licked an ant's butt, and I had provided nourishing blood for the local mosquitoes and leeches. I had melted in the heat and been numbed by the creek, been thrown off balance in eons-old mud and disoriented by echoing frog song. I had become one with the rainforest!

The next morning, as I counted the more than seventy-five bites on my lower left leg alone, Jim laughed and said, "Let's do it again!" And I knew he meant it—because, in the end, this was exactly the adventure we had come for.

Unique Species

No boys allowed: The desert grassland whiptail lizard is a female-only species.

She-Lizards of the Chiricahua Desert

Sexual renegades in the Arizona wilderness

Hidden in remote locations among the towering buttes and high-desert scrub of southeastern Arizona lives a remarkable population of parthenogenetic lizards. For those who haven't memorized the details of rogue reproductive biology, parthenogenesis is the ability of females to reproduce asexually—no males are involved. In fact, these parthenogenetic lizards are *all females*; there is no such thing as a male of the species. I'm intrigued by the idea of such sexual renegades—who wouldn't be?

Biologists have long known that certain invertebrates, such as bees, wasps, and aphids, depend on parthenogenesis. Komodo dragons do it regularly though not exclusively; their embryos develop just fine from unfertilized eggs. Crocodiles and several species of sharks resort to it on rare occasions.

The desert grassland whiptail lizard *(Aspidoscelis uniparens)* is famous for her parthenogenetic abilities. She's a pretty little thing, just about five inches long, plus her tail. And what a tail it is! Slender and a bit longer than the lizard's body, it often lies in a casual-yet-elegant curve as she sunbathes on a rock. In hatchlings, it's a particularly fetching bright blue. Her scaly coat is a sophisticated buffy-olive color, and six pale-yellow stripes running down her back—sleek and sporty—emphasize the she-lizard's lithe figure.

All in all, desert whiptails are quite an enchanting species; I was eager to see them. One spring, I traveled to Arizona's Chiricahua Wilderness Area (pronounced "cheer-a-COW-a") to have a look.

It wasn't easy to get there. From northern California you have to drive hundreds of miles south, then cut over and cross Arizona on Interstate 10, exiting at Roadforks, New Mexico. From there you head south and zigzag back into Arizona. Next you take a treacherous, barely paved road with soft shoulders for quite a way, continue onto a dirt road, and cross a rugged wash. That last part you can only do in a four-wheel drive.

The name of the place refers to the Chiricahua Apache, nomadic people who lived here and were led by renowned chiefs—Cochise and Geronimo among them. The area is bounded by the Sonoran Desert to the west, the Chihuahuan Desert to the east, and the mountains of northern Mexico on the south. Yucca and scrubby native grasses dot the landscape. The air is thin; the sun burns hard and bright.

Plants and animals have developed adaptations for survival in this harsh climate. The prickly pear cactus, for example, retains water in its fleshy body. Instead of traditional leaves, it has developed sharp spines as protection from marauding animals in search of moisture. Another example: The yucca leaves' waxy coating helps prevent water loss.

Kangaroo rats use a similar survival strategy; their super-oily coats help conserve water. That's just the beginning of their adaptation for desert life, though. These little rodents don't get much of anything to drink, so they effectively recycle their own urine to conserve water. They also have extra-convoluted nasal passages that act as little heat exchangers, cooling their breath so less moisture is lost when they exhale.

Gila monsters and chuckwallas have their own methods: They each store water in fatty lumps—like camels do, except the reptiles' lumps are in their tails. Horned lizards have an intricate system for capturing dewdrops on their bodies, and spadefoot toads can go into a state of suspended animation for months at a time when food or water is scarce. But as we now know, these feats of desert adaptation are nothing compared with those of the she-lizards.

A quick informational detour is in order here to explain parthenogenesis. I'm not a scientist, but this is what I understand: In certain populations males are absent or in short supply, often for reasons that have to do with habitat availability. So the females just go ahead and lay their eggs without the services of a male or even sperm, and without any of the muss and fuss that typically accompany sexual congress.

Actually, that is not completely true. There *is* some muss: an unusual chromosomal-triplet-based reproductive method called triploid polyploidy*—a trait the whiptails share with tardigrades and seedless watermelons, and which is too complicated for me to understand.

There is also a bit of fuss: Female whiptails have been observed mounting other female whiptails, in what the scientific literature describes as "female-female courtship and mating rituals." The encounter apparently stimulates ovulation in both participants. At any rate, this almost unbelievable method of reproduction allows a single female to establish an entire new colony, expanding the range of the species.

So if, for example, a roadrunner, which is a predator of the whiptail lizard, were to snatch up a whiptail and

* Because of the way triploid polyploidy works, the offspring would not be *exact* clones of their parents, but they'd be very, very close.

fly some distance away with it, but then accidentally drop it, and the whiptail fell, remarkably unharmed, to its freedom in a new-but-whiptail-less land, it would not mean the end of that lizard's genetic line. It would instead be the beginning of a brand-new population, and it must have seemed like a splendid evolutionary development ... at first.

But there *is* the matter of genetic diversity. All the genetic mixing involved in "regular" sexual reproduction is why we are each unique, and why our species, made up of diverse individuals, is genetically healthy. The lizards' parthenogenetic reproduction is tough for the community, because without diversity it's more susceptible to disease and adverse environmental changes.

It must be tough for the individual lizards, too. Think of it: With no genetic diversity, every lizard would essentially (triploid polyploidy) be a clone of her mother. Same tiny ears, slippery scales, and long toes as her mother had. Not only that, but the mother lizard would be a clone of *her* mother, and so on, back many generations. There'd be little opportunity for variation or evolution—longer tails, keener eyesight, improved camouflage—as would be possible in a population that reproduced sexually.

Imagine what it would be like if we humans had to bear the weight of that endless monotony: identical women stretching out in every direction, even extending backward and forward in time. Mothers, daughters, sisters, aunties, grannies—everywhere you looked, you would see yourself, as though you were living in a hall of mirrors. Glance in one direction and you'd be treated to a vision of your younger self: girlish and fresh. Turn another way and you'd see a specter of your future self: older, slower, wrinklier. Time would dissolve; identity would shift. What would "self" even mean in a colony of clones? In some weird, psychosis-inducing twist you might even glimpse an eternity of universal oneness. It would be the stuff of nightmares.

This self-cloning behavior seemed to me like more than just a fascinating curiosity of the animal kingdom; it seemed, potentially, like an existentially *big deal*. Or, at the very least, some kind of cosmic metaphor. At any rate, it's why I really wanted to see the lizards, and why I traveled all the way to a wilderness area and dragged myself through a desert of netleaf oak, yucca, and fan palms, punctuated by the occasional prickly cholla cactus, for a glimpse.

The March sky was bright blue, the puffy white clouds picture-perfect, the breezes redolent with juniper and pine sap. My trip was expertly timed—the whiptails' hibernation season had just ended—so I expected to see loads of them basking in the sunshine, and maybe a few hiding near small seeps or springs, stealthily awaiting their prey.

They are excellent hunters, by the way. Whiptails are so strong they can catch and eat reptiles, mammals, and birds that are bigger than they are. And they can apparently detect underground predators or prey—from above ground. Interestingly, science has shown that desert whiptails are measurably less aggressive than other lizards. These remarkable creatures are such an enigma! I had a sort of science-crush on the whip-tails, and could hardly wait to spend quality time hanging out with them.

At first I didn't see any, so I figured they must all be perfectly camouflaged on the Chiricahua's lichen-encrusted outcroppings. I hiked up higher to have a look, scanning each sunny rock, every shadowed crevice. Rocks crumbled into slippery gravel beneath my boots, trails disintegrated, and the mile-high air slowed me down, but I kept up my search for an entire week.

However, it was unseasonably cool the spring I visited—a full ten degrees below average—and despite my best efforts and all my positive visualizations, the creatures were nowhere to be found. Apparently, they were still underground.

So, I never did get to see a desert whiptail—but I did leave the Chiricahua inspired. In a place that tests endurance at every turn, these sexual soloists are giving up genetic diversity in a bid to avoid extinction. Their story is a testament to adaptability, evolution, and the wild ingenuity that allows life to persist in the most unforgiving places.

A greater bamboo lemur,
ready to decimate a patch of bamboo

Lemurs and Leeches

Field notes from a vanishing Eden

I was ducking into the camp outhouse when an urgent whisper cut through the forest: "Bamboo lemur, behind the loo!" I made a judgment call. The greater bamboo lemur *(Prolemur simus)* is one of the rarest primates in the world; researchers believe there are fewer than 500 individuals left. In fact, this critically endangered species was believed to be extinct until it was rediscovered only a few years ago. Never mind the loo—I rushed out, hoping for a quick look.

The lemurs had teddy-bear-like faces, round chestnut-colored eyes, and fluffy, pale grey ear tufts that made them look like little Yodas. Several came right down to eye level and sat only an arm's length away, watching us nonchalantly as they broke giant bamboo stalks, stripping off the tough outer layers and munching the inner pith. Greater bamboo lemurs have adapted to process cyanide-laced bamboo, and thus occupy a rare and peculiar ecological niche.

This was nothing like seeing animals in captivity—through glass or bars, or across a wide moat—wondering about their adaptation to zoo trees and zoo food and zoo neighbors. This was the real thing! I could have reached out and touched their soft gray fur, or rubbed them behind the ears like my cat. I might have tried, too, except that teeth made for eating bamboo clearly mean business, and these were wild animals. I watched for more than an hour, my heart pounding with excitement as five lemurs moved methodically from one stout stalk to the next, wreaking havoc on the bamboo grove. This species requires a large territory.

I was in Madagascar with a group of Earthwatchers who had volunteered to help Dr. Patricia Wright collect field data for a study of lemurs, specifically *Propithecus diadema edwardsi* at Ranomafana National Park. Perhaps you know *Propithecus* by its common name, the Milne-Edwards' sifaka. Well, perhaps not. But surely you're familiar with lemurs, those cute, monkey-like prosimians found only in Madagascar. Like almost all plants and animals there, lemurs are in danger of extinction; conservation efforts during the next few years will be critical for their survival.

Dr. Wright—whose interest in primate research was whetted by a pair of pet Peruvian monkeys she owned as a young housewife in Brooklyn—has been studying lemurs since 1984. Her vision and hard work were largely responsible for the establishment of Ranomafana National Park, and she was awarded both the MacArthur "genius" grant and Madagascar's Chevalier d'Ordre National (the equivalent of knighthood) for her efforts there. I felt honored to be working with Dr. Wright.

Honored and smelly. Gary, Pam, Christine, and I ("Team 3") ran through the rainforest together, sweating profusely, for two weeks. Our legs grew strong, as did our odor. We had to move fast to keep up with the lemurs; every day they led us through the forest, off the trail, down steep, vine-covered slopes, across a tributary of the Namorona River, along the rocky bank, back across the river—several times—through the tangled undergrowth, and back up the river's slippery slopes. I couldn't help but think they were toying with us.

Not only that, but the rainforest was teeming with stinging caterpillars, malaria-carrying mosquitoes, biting horseflies, and blood-sucking leeches. Spiders the size of my fist were common, and I was regularly dive-bombed at dinner by moths bigger than a hummingbird.

When we first arrived at the jungle outpost, I noticed that some of our guides smelled pretty ripe. I figured on keeping my distance until I got used to it. Little did I know that I—and all my fellow Earthwatchers who had arrived fresh, clean, and eager —would soon smell of sweat and mildew and the boiled zebu meat that was our staple dinner.

Twelve of us, strangers from around the world, had volunteered two weeks of our precious vacation time to track lemurs in the service of environmental science. Our group included a pharmacist, a private pilot, a dentist, two bored housewives, a member of the British Royal Air Force, several avid birders, and me—a nature-lover since childhood—plus a handful of graduate students working on research projects. I fantasized that the pharmacist's large backpack transformed, with the push of a button, into a James Bond-issue collapsible stretcher, which would be useful if I slipped on the wet trails, tumbled down the mountain, caught my foot in a tough vine, and broke my leg. The pilot could evacuate me on the collapsible stretcher in an emergency, although we would need to find an airplane.

Even though we were a varied bunch, our reasons for participating were the same: we all cared deeply about the environment, and saw this as a way to express our commitment with action. Several of us (including myself) were at midlife transitions and wanted—needed—to do something that would have a real impact on the world. And we were each hungry for adventure.

Why *this* adventure? Because Madagascar can't wait. The land is starving, dying. The people of Madagascar have traditionally subsisted with slash-and-burn farming: cultivating a piece of land for a few seasons until the soil's fertility is exhausted, then abandoning that plot, moving on, allowing the earth to regenerate. But the people have run out of land, and the land has run out of time. Whole ecosystems are dying, and species are becoming extinct. We were there to save the world.

Our headquarters for this endeavor was a one-room log cabin. It was packed with an assortment of blue plastic tarps and leaky tents, paperback travel stories in several languages, hiking boots that fit everyone and no one, precarious stacks of field data books, and damp, patient messages awaiting researchers who were somewhere deep in the forest. There was also a pantry, stocked with delicious-smelling local honey, tins of sweetened condensed milk, and a few recycled jars filled with sugary peanut butter, generously donated by the wife of one of our Malagasy guides. We slept in the leaky tents.

Although we had been warned to expect pit toilets,

we found two semi-private outhouses, with toilets that flushed most of the time. The only other permanent structures in camp were four shower stalls, in which hundreds of tiny gray spiders had taken up residence. We used these facilities with a sort of grateful trepidation.

After the lemur-by-the-outhouse incident, I discovered a small black leech attached to my right hip. I must have got it in the outhouse and not noticed because of my rush to see the greater bamboo lemur. Anyway, I brushed it off—with minimal bloodshed—and quickly warned the rest of the group. Others shared their own leech stories—encounters involving noses, mouths, and other places best left undescribed.

These rainforest leeches are small and terrestrial, which means you're never safe from them, even at night, all zipped up inside your sleeping bag and tent. But I found that with leeches—like many things—the anticipation is far worse than the actual experience.

Madagascar today is a conservation hotspot; most of the rainforest is gone, and many of the unique species here are in imminent danger of extinction. Although international money is flowing into the country, not enough finds its way to the local people, and centuries-old patterns of *tavy* (slash-and-burn farming) are almost impossible to change. In a subsistence-level economy with inadequate health care and education, it's difficult to convince people to switch to sustainable agriculture. The little rainforest that remains is under tremendous pressure, since the rapidly growing Malagasy population needs land to produce food.

The lemurs we followed are almost totally arboreal, eating, sleeping, and traveling in the treetops. If trees are removed—either by selective logging or by clear-cutting—the resulting gaps in the forest canopy restrict the lemurs' territory, and their movement within their territory. This is true even if primary forest has been replaced by secondary growth; it's old-growth trees—ones that have been around for hundreds of years—that are intertwined with the vines and lianas so important for arboreal locomotion. Forest degradation is also a problem because these tree-dwelling animals don't have adaptive strategies for dealing with predators on the ground.

My job was to collect data. Every ten minutes, I used my Rite-in-the-Rain pen to record an adult female lemur's behavior in a waterproof field data book. What was she doing: feeding, grooming, resting, or traveling? What kind of tree was she in, and how high? If she was feeding, was it on fruit, seeds, flowers, or leaves? I also recorded the focal lemur's nearest (lemur) neighbor, and distance to nearest neighbor, which will eventually help researchers understand more about the relationship between diet and social structure.

Our Malagasy guides were essential; they were experts at tracking lemurs as they leapt from vine to vine and scampered up and down trees. They knew the names of all the plants and were good at estimating height. (Whereas I found myself wondering how many men stacked on top of one another would equal the height of the lemurs' treetop feeding spot—five tall men? six?—and then doubling the number to convert to meters.)

My guide was Dominique, a short man with shy eyes who moved quickly and easily through the rainforest. He seemed completely at home there, and very different from us volunteers—foreigners, loud and awkward in the undergrowth. Following along right after Dominique, I imagined that I picked up some of his gracefulness and agility; it was so easy to walk with him through the forest!

Later, I realized it was easy because Dominique discreetly pushed lianas out of my way, pointed out thorns and treacherous holes in the forest floor, and subtly sliced the occasional threatening vine with a pocketknife hidden in his hand. We didn't speak the same language, but Dominique's actions communicated his generous nature. When we crossed the river, he balanced carefully on dry stones so I'd know where to step, even though he was wearing waterproof rubber boots. (I was not.)

Why bring foreigners into the country to track lemurs? Although the local Malagasy people know the rainforest very well, they have a decidedly non-Western approach to data collection. Our conceptions of time, accuracy, consistency, and detail are not theirs.

And Dr. Wright needs to collect a lot of data. The more researchers understand about lemurs' lives, the more successful they are likely to be in figuring out how humans and lemurs can live together. Also, by focusing attention on lemurs and other endangered species, we encourage the Malagasy people to preserve the remaining rainforest.

Most of the island's wide range of plants and animals—more than 200,000 species—have evolved in isolation from the rest of the world. If you don't count birds, which can fly there from mainland Africa, more than 80% of Madagascar's plants and animals are endemic; they are found nowhere else on Earth.

Madagascar's evolutionary history is complicated and paradoxical. How can it be that this island, only 270 miles off the eastern coast of Africa, shares none of the continent's large mammals, such as elephants, antelopes, or lions? Or that it has only one insectivore, the tenrec, a hedgehog-like creature whose nearest relative is native to Cuba and Haiti?

Collecting field data for Dr. Wright was not easy. I was always hot and tired, sweating and thirsty, and desperate for a shower. My glasses fogged up during the daytime, and every item of clothing I took ended up smelling—either vaguely or distinctly—of mildew. I paid good money to spend two weeks of my life chasing lemurs and recording their most mundane behaviors.

The female tailless common tenrec, by the way, can produce up to 32 offspring at a time—the record for mammals—and individuals have been recorded with up to 29 nipples. Tenrecs' super fecundity is more than an interesting curiosity, though. It helps explain the wild speciation they have undergone on the island. Lemurs, chameleons, finch-like vangas, and palm trees have also produced far more variety than taxonomists would expect in such a small area.

But I have fallen in love with the rainforest. I love the souimanga sunbird's high warble, the insistent "welcome sunshine, come sunshine" of the cuckoo, the rushing Namorona River, the greenness of the canopy overhead, and the tangle of vines underfoot. I am fascinated by the unique wildlife and unearthly landscapes in Madagascar. And I have discovered that even blood-sucking leeches aren't so bad, after the initial getting-to-know-you phase of the relationship.

I volunteered because the lemurs are endangered, and collecting ecological data is an important first step in helping to preserve their habitat. I'll return to help again because rainforests are the Earth's lungs, and my grandchildren will need to breathe.

This is not the same kiwi I saw snuffling through the woods at Zealandia. This one happens to be dead and stuffed, but it still appears somewhat lifelike if you squint a little. Like I said: a roving hairball.

No Moa, No Moa

In praise of flightless birds

Hunting under cover of darkness, I'm at a disadvantage: A red-lensed flashlight is my only source of illumination. Even that may be too much, as the creature I'm after is shy and nocturnal. I hear rustling in the dry leaves and quick scurries off the path, but I haven't glimpsed one in the hour I've been searching.

This expedition is a big deal for me. I'm halfway around the world—in Aotearoa—tracking down a rare bird. Aotearoa is the native M ori name for New Zealand. Its literal translation is "Land of the Long White Cloud," and the name resonates with a particular sadness I feel about the country. Or more specifically, a sadness I feel about the birds that once lived here, for theirs is a sorry story: Many have gone extinct since the arrival of humans. A giant goose, the fierce Haast's eagle, the odd-looking adzebill, and a fascinating bird called the moa—all once thrived in New Zealand but have since died out.

The bird I'm searching for has not disappeared, however. It's called a kiwi, and it is such a strange creature that I decided I must see one with my own eyes. The kiwi lives only in New Zealand and is flightless, for starters. It has no apparent wings or tail, and its speckled plumage looks like hair rather than feathers. It is also nocturnal, has a highly developed sense of smell, and sports nostrils at the far end of its long, downward-curved bill—all unusual in a bird.

And then there's the egg. The kiwi lays an egg that is proportionately larger, compared with the size of the mother, than that of any other bird alive; the eggs can be up to one-quarter the weight of the female herself. This is an incredible feat; think for a moment of your own weight, and then consider birthing something that weighs one quarter of that amount. Enough said.

These impressive peculiarities pushed the kiwi toward the top of my must-see list, so once I got to New Zealand, "See a kiwi" was the first item on my itinerary.

To clarify a sometimes-confusing term, the kiwi bird—the critter I wanted to see—is a national symbol for New Zealand, which is why people from New

Zealand are commonly called Kiwis. The fruit we call a kiwi is also called a Chinese gooseberry, and is native to China. It is so named because it has been commercially grown in New Zealand, renamed as part of a marketing effort, and exported to the United States as "kiwifruit." I can't speak to the origin of Kiwi shoe polish, which has a picture of the kiwi bird on the packaging, but was not developed in New Zealand.

But back to the bird: There *is* a way to see the creature—rare, shy, and nocturnal though it is. Ten minutes from the center of New Zealand's capital city, Wellington, sits a remarkable sanctuary called Zealandia. Its 550 wooded acres are protected from introduced predators by a state-of-the-art fence that completely encircles the grounds and extends eighteen inches below ground level to keep out digging animals.

Security above ground is tough, too. When I visited one drizzly April afternoon I had to stop and open my daypack to confirm it contained no small freeloading fauna. Then I passed through a series of gates, each of which had to be closed before the next would open. They're serious about keeping out dogs, cats, ferrets, and other non-native predators that could decimate the population.

Thus protected, some forty species of native birds and dozens of reptile species flourish here. Many are endangered, including a small population of little spotted kiwi *(Apteryx owenii)*. It's been extinct on New Zealand's mainland since 1875, but a few individuals were recovered from a small island, bred in captivity, and released in the sanctuary in 2000. In the wild, only one in twenty chicks survives, but here at Zealandia they do a little better.

The little spotted kiwi is the bird I'm here to find, the reason I'm traipsing through the dark, damp forest with a young Zealandia guide named Sara. Heavy clouds cover the moon tonight, and without our red flashlights I cannot even see the path at my feet.

Sara tells me more about the kiwi. "New Zealand evolved without any mammals," she begins.

"Really?" I am doubtful. The islands, about the size of California, surely could not have developed without a whole class of animals. Nearby Australia, with its crazy collection of critters like the kangaroo, dingo, and duck-billed platypus, has lots of mammals.

"OK; you got me. There were three species of bats. That left a large ecological niche open, so birds evolved in New Zealand much as mammals evolved in other places—right to the top of the food chain."

"So the birds became predators?"

"Yes, but what's even more interesting is that because there were no mammals to prey on them, many of the birds, including the kiwi, gradually lost the ability to fly. Without the capacity for flight, they no longer needed the lightweight bones typical of birds, and developed heavier ones that allowed the females to lay very large eggs."

That isn't to say it was easy. Producing a huge egg, I learned, places a lot of physical demands on the female kiwi. For the thirty days or so it takes to grow the fully developed egg, she must eat three times her normal amount of food—mostly worms and other invertebrates, a few berries, maybe even a small frog. She becomes so large that very little space is left inside for her stomach, forcing her to fast for the last several days before the egg is laid. The male incubates the egg—which seems only fair, given what the female has to go through.

Understandably, only one egg is generally laid per season. That worked out fine for a long time, as the kiwi had few predators. Then came the humans. Arriving from Polynesia less than eight hundred years ago, they quickly exploited the island's abundant large game, which consisted mostly of birds.

One of the first to go was the giant native goose *(Cnemiornis)*, which stood about a meter tall and weighed nearly forty pounds. It laid giant eggs—as one would expect a giant goose to do—and giant goose eggs were doubtless a nutritious staple of the early islanders' diets. The giant goose was extinct by the turn of the fifteenth century, barely one hundred years later.

The yard-long, tiny-winged, flightless adzebill—probably so named because its huge, heavy bill resembled an adze or an axe—lasted until the beginning of the sixteenth century. It also weighed nearly forty pounds, and would have made a good-sized meal.

The moa lasted a little longer. About a dozen species, all flightless, roamed New Zealand's forests, grasslands, and mountains. The smallest of them was the size of a turkey, but the largest—well, the largest moa was the biggest bird that ever lived. It was more than ten feet tall and weighed as much as five hundred pounds.

Some think moa were still alive at the beginning of the seventeenth century, and an actual mummified moa neck was found in Earnscleugh Cave on New Zealand's South Island in 1871. The famous "Earnscleugh Neck" moa mummy is a sixteen-inch-long slab of meat, skin, and bristles which, as one wag put it, "looks for all intents and purposes like a large pork roast that was overcooked and hidden in a dry cave for two millennia." Despite that awful description, the moa has become a symbol of extinction, immortalized in a poem sung by New Zealand schoolchildren:

> No moa
> No moa
> In old Ao-tea-roa
> Can't get 'em
> They've eat 'em
> They're gone and there ain't no moa
> —Traditional New Zealand song

The flightless moa, unable to defend itself from humans, may have been destined for early extinction. But ironically, the unadaptive bird evolved in close association with a spectacularly adaptive species of plant. The lancewood tree *(Pseudopanax crassifolius)*, which still grows in New Zealand's lowland forests, "learned" to survive alongside legions of hungry moa that could easily devour a sapling in a single sitting. The tree has evolved to change its appearance as it ages, its varying morphologies protected from birds in different ways at different times in its lifecycle.

As a seedling, the lancewood is a mottled brown color and blends into the forest floor in what seems like an attempt at camouflage—it looks a lot like the leaf litter it sprouts up from, and would not have been easy for a marauding moa to spot.

Once it is too tall to hide anymore, the lancewood goes on the offensive. Its leaves become weaponized: They turn into rigid, yard-long, downward-pointing daggers, often with sharp barbs around the edges like the snout of a sawfish. These vicious leaves would have been perfect for fending off a browsing bird.

I find all this adaptation quite impressive, but the lancewood has yet another trick up its sleeve. When it reaches about three meters high—which, by no coincidence, was the height of the largest moa—it changes again. Its leaves lose their barbs and turn upward toward the sun. Atop a tall, straight trunk, they would have been beyond the reach of a hungry moa.

The radical transformations in the lancewood's shape and color are fascinating, but it's those ancient giant birds that really captured my imagination. To my mind, the most impressive was Haast's eagle, the largest eagle the world has ever known. It had a foot three times the size of a modern-day bald eagle's, its talons were the size of a tiger's claws, and its predatory power must have been phenomenal. This eagle would dive on its prey, such as the land-bound moa, at a speed of up to fifty miles an hour, piercing and crushing the flightless bird's neck with its mighty talons. Not surprisingly, the Haast's eagle died out along with the extinction of its most important food source: No moa, no eagle.

Learning about the extinctions of these incredible birds leaves me feeling sad. Any extinction is an unhappy event. It seems especially poignant, though, when the species in question is superlative: the biggest goose, the tallest bird, the mightiest eagle the world has ever known. The *only* adzebills.

But there is good news. New Zealand's kiwis are still alive. This may be in part because, except for the chiefs, M ori law forbade eating kiwi. To kill one, even accidentally, was to court death.

This law might have been enacted because of those giant eggs, a wonderfully renewable source of food. Or perhaps it was simply because of the taste of kiwi meat, which one European settler described as gamey and tough, adding, "Imagine a piece of pork boiled in an old coffin."

"It isn't mating season yet," Sara says, "but we've already been hearing pairs calling to each other. The male will start; his voice is higher and shriller. Then the female joins in with a harsher, croaky kind of call."

A minute later, we hear one. It's a male, Sara says, about twenty feet off the path, and effectively invisible to me. She catches him with her red flashlight and we watch as he wanders through the brush, moving quickly, but not overly concerned by our presence. He doesn't look like a bird—more like a roving hairball, the size of a chicken, with a long, slender beak that he pokes into the leaves on the ground.

Before I can ask what he's doing, Sara explains in a whisper, "Their nostrils are at the far end of their bills so the kiwi can sniff out underground insects and worms in the dark."

And then he's calling again, sixteen times in a row! The sound begins like an ill-tempered toddler protesting some moderate injustice, then warbles and rises to an emphatic question at the end. A lower voice joins in from a short distance away, and they finish out the call-and-response.

"Once they're bonded," Sara adds, "a male and female kiwi usually live the rest of their lives as a monogamous couple. They call out to each other at night during the mating season; that's what we're hearing tonight." We listen a while longer, but the two kiwi have gone—snuffling off through the forest together.

"They can live up to twenty years," whispers Sara. That seems like a long time for a bird, especially given the girth of those annual eggs, but in the larger scheme of things it's only a heartbeat, the blink of an eye ... the flap of an ancestral wing.

A quiet relief settles over me. In a country that has lost the biggest goose, the mightiest eagle, and the tallest bird the world has ever known, these small, flightless birds endure. They rustle and call in the darkness—ghosts with feathers. And here, under the long white cloud, I am witness to their resilience, a part of their moment, if only for a heartbeat.

Oddest of all is its third eye...

The Third Eye of the Tuatara

An ancient cannibal skitters toward extinction

On a small island in the South Pacific another survivor moves through the dark. It is older than most birds, older than mammals—a cannibal with a crest of spiky spines down the center of its back and tail, three rows of teeth, and invisible ears. This strange creature starts out small, but keeps growing larger and larger for the first thirty-five years of its hundred-year life. Scientists have studied its genome, which suggests it is a freakish amalgamation of lizard, bird, and mammal. Oddest of all is its third eye.

This lizard-like reptile, called the tuatara, survives in the wild only in New Zealand. I happened to be headed there for a conference, and as soon as I heard about the tuatara I was determined to see one.

"We have to figure out a way to see a tuatara while

we're in New Zealand," I announced to my husband, Jim, who planned to join me on the trip. "I've done some research."

"I'm sure you have," he replied, peering at me over the top of his glasses. "Let's start with the scientific name." (He knows me really well.)

"*Sphenodon punctatus*. It's a species like no other. Actually, it's a genus, a family, and an order like no other, because it's the sole survivor of the *Rhynchocephalia/Sphenodontia* order."

"Your point, Laurie?"

"It's ancient! The tuatara predates even the earliest dinosaurs. All of its closest relatives are fossils."

"So we won't be meeting the whole family then. Is this another one of your wild goose chases ... like the cassowary?"

Ah yes, the cassowary. At my insistence we'd gone on an Australian cassowary hunt and hadn't seen a single one. I wasn't sure we'd see a tuatara either, since it is so rare and lives, in the wild, only on a few small islands off New Zealand's mainland. But viewing one was worth a try. I'd been obsessing about the creature, and my research had turned up all kinds of fascinating facts.

Most sensational of these was the story of Henry, a tuatara that became a surprise father at the age of 111 "after receiving treatment for a cancer that had made him hostile toward prospective mates." The UK-based *MailOnline* went on to say that Henry was "caught canoodling with a female named Mildred" and that the pair had produced eleven offspring. An accompanying photo showed Henry and Mildred *in flagrante*.

A casual observer could be forgiven for thinking a tuatara is a lizard. It *looks* like a lizard. Like true lizards it has scales, lays eggs, and eats insects and other invertebrates. But there are significant, if inconspicuous, differences: that third row of teeth, for example. The tuatara's row of bottom teeth fits neatly in between two parallel rows of top choppers, which the tuatara moves in a slicing motion rather than up and down. This gives even young tuatara the ability to rip apart hard-bodied insects. It makes the larger adults fearsome predators that can decapitate birds, and sometimes even eat their own offspring.

Also unlike lizards, tuatara are nocturnal. This may be because they function best in cool weather (lizards like it warm). In fact, the tuatara has an extremely low optimal body temperature—the lowest of any reptile—and can remain active in near-freezing temperatures.

And then there's my favorite part, the third eye. Situated on top of the tuatara's head and more primitive than a regular eye, it is barely visible. Still, it has a lens, retina, and nerve endings, and is sensitive to light. It turns out that many species of lizards, frogs, and salamanders also have this "parietal eye"; lampreys have two. The eye probably helps animals with navigation, thermoregulation, and managing circadian rhythms.

I've long been fascinated with the idea of a third eye—as has much of the rest of humankind. In fact, we've linked this enigmatic aperture with the pineal gland and assigned it special importance since ancient times. In Pharaonic Egypt, beginning in the thirty-second century BCE, priests associated the pineal gland with the Eye of Horus, a symbol of power, protection, and good health. Hindu yogis connected it with Sahasrara, the chakra of enlightenment. Second-century Greeks believed the pineal gland regulated "psychic pneuma," which was "the first instrument of the soul." Seventeenth-century philosopher René Descartes declared it was the source of all thought, as well as the "principal seat of the soul."

In current times, New Age body-mind practitioners like Dr. Joe Dispenza — whose training includes neuroplasticity, epigenetics, and brain/heart coherence—have written about the pineal. Dispenza pointed out in 2017 that if you apply Fibonacci's constant to a cross-section of the human

Dr. Dispenza describes a method of coordinated breathing and muscle control that he says induces a wave in the cerebrospinal fluid, activating the pineal in a way that primes the brain for mystical experience.

brain, following the resulting spiral along the circumference of that organ, it will end precisely at the pineal gland—signifying, at the very least, that the pineal must be something quite special.

The tuatara fascinates scientists for other reasons, as well. It is uncommonly resistant to infectious diseases. It has an unusually large number of genes involved in protection against cellular aging, which might eventually provide insights that help humans live longer. And X-ray microtomography scans of particular small tuatara bones and cartilage formations are helping morphologists understand the evolution of kneecaps.

One would expect such a remarkable species to be held in great esteem, and in fact it is, by the M ori people, who view the tuatara as a guardian of knowledge and regard it as a *taonga* or special treasure. But despite that esteem—as well as the efforts of New Zealand's Department of Conservation—habitat loss and predation by rats and other introduced pests have led to the tuatara's declining population. Which is why Jim and I could only see the tuatara at a zoological park near Wellington, rather than in the wild. Additionally, the tuatara, being nocturnal and very sensitive to light, is not viewable during daylight hours.

"We're going on a night walk," I announced.

"Not another one."

"They're fun. Remember the kiwi?"

"What kiwi?" Like the cassowary hunt, our kiwi night walk had been a bust. It had taken several hours, and although we'd heard the bird calling, we saw only its legs and a flash of its hair-like feathers. This promised to be a similar situation, but the strength of my determination won the debate. We had to hang out at a nearly empty wildlife park for several hours after closing time, waiting for darkness to fall. The night was moonless and cold, Jim's feet were hurting, and boredom had set in. So had irritability. "This had better be worth it," he warned.

When we finally set out, we had to use red-lensed flashlights so as not to disturb the vision of all the nocturnal creatures we were sure to see. Those creatures turned out to be mostly glowworms, which, although magical in their own right, were not the three-eyed cannibals I longed to encounter.

As we walked along the narrow path, deeper and deeper into the forest, dry leaves crackled beneath our feet. A cool breeze gave me the shivers. Jim was right—this had better be worth the trip. I turned to him. "I want to point out that the tuatara is an amazing creature. It's one of the world's longest-living and slowest-growing animals. In fact, it's a national icon of slowness."

"If it's so slow, why did they name a race car after it?"

"I don't remember reading anything about race cars. I don't think that's right."

"Yes it is. I've done my research, too. Shelby makes a car called the Tuatara. It's got a top speed of nearly 300 miles an hour."

"Hmmm. A testosterone lover's dream," I mused, poking the bushes with my walking stick. "Speaking of which, did you know that a recent comparison of tuatara embryos with fossil evidence suggests that external male genitalia originated just once in amniotes?" I wasn't sure where I was going with this information, but it sounded like a promising line of conversation.

"Amniotes? What are those?"

"All mammals, birds, lizards, crocodiles ... and the tuatara. But external genitalia have since undergone dramatic modification, disappearing altogether in birds and reptiles. There's more. The gender of tuatara, like that of some tortoises, is determined by the temperature of its nest. When incubation temperatures rise above 72°F, tuatara hatchlings all turn out to be males."

"Good if you're a female tuatara," Jim says, raising his left eyebrow suggestively.

"Not so good for the species, though," I counter. I

hate it when he wiggles his eyebrow. It makes me want to argue with whatever he has just said. I get right to it. "Climate change is going to be a disaster for them."

As it turned out, we found little to amaze us on that three-hour exploration under the cloak of darkness. No triple rows of teeth. No mysterious third eyes. No dragon-like creatures devouring their own young. What we *were* able to do was watch as three small lizards—ok, tuatara—skittered across the dark path in front of us into the shelter of nearby bushes. All were under a foot long—including the tails—and had minuscule rows of knobby spines decorating their necks, backs, and tails. They had big eyes that glowed red in the light we provided, and large, expressive forefeet.

And that was the end of our adventure.

I feel sorry for the tuatara. Once widespread, there are now fewer than 80,000 individuals. The youngsters must be wary of being eaten by their parents. They also have the constant problem of attack by rats, and they don't even have real teeth. Instead, their jaws are like saws, with serrations poking out of the jawbone. If the "teeth" break off or are damaged, there's no way for the tuatara to grow them back. And, with their one-hundred-year lifespans, this sad situation is a long one. So, all in all, not the most promising outcome for the bearer of an impressive and mystically celebrated third eye, the last survivor of a genus, family, and order.

On the bright side, though, there's Henry—still canoodling at 111.

Life on the Edge

Stravinsky, the fiery-billed aracari, from his good side

Stravinsky's Gift

A one-legged toucan offers a surprising gift
in the Land of Lucky Mud

Stravinsky stretches languidly, neck to toe, then falls onto his butt. With only one useful leg and no tail feathers to speak of, his balance is compromised. He cannot perch. He is forced to live on the floor of his cage, hopping on one foot, tail feathers worn to nubs. The flight feathers on each glossy green wing—which should be long and elegant—are also shortened from constant rubbing against the floor. They give this small toucan the sad, stunted look of a multiple amputee.

Peering into Stravinsky's cage, I am immediately drawn to him, although I cannot say why. But Stravinsky will not make eye contact—his only acknowledgement of my presence is a nervous hop to the far side of his cage. I wish the skittish creature would reciprocate my interest.

We have come together at Finca Fango de la Suerte, the sprawling central Costa Rican home of Joan Hall and Lenny Karpman, who have created a sanctuary here for more than forty injured and unwanted animals. The climate is nearly perfect, the air is clear, and the *finca* exudes a warm, amiable hospitality. Stravinsky has lived here for three years, recuperating from his wounds. I am visiting for two weeks, craving sanctuary myself.

I love the name of the place: A *finca* is a small ranch, and *fango de la suerte* translates roughly as *lucky mud*. It's a reference to *Cat's Cradle*, the Kurt Vonnegut, Jr. novel in which God created the earth and then woke up the mud so it could see what an excellent job he had done. "And I was some of the mud that got to sit up and look around. Lucky me, lucky mud ... I loved everything I saw!"

Joan and Lenny love everything they see, too. They have brought homeless cats, dogs, and birds to Finca Fango de la Suerte. The place is a tropical madhouse. Roosters crow at midnight and kittens eat butter off the breakfast table. Parrots laugh raucously and screech "Grandpa! Grandpa!" when Lenny walks by. Six dogs lick my feet and legs at every opportunity. Their tongues are long and soft.

But it is Stravinsky, a fiery-billed aracari, who captivates me. A riff of red across the small toucan's golden breast conjures blood, as though he'd recently lost a fight. Stravinsky *won* a fight, though—a fierce one. Years ago, he was attacked as a nestling by a marauding rat, which gnawed all the toes off the tiny bird's right foot after brutally devouring his sibling and nest mate in its entirety. Left alone, Stravinsky would have starved to death—if he had not been eaten first by a snake, a coati, or another rat. But Stravinsky was saved and brought to Amigos de las Aves, a rambling avian rescue center founded by Richard and Margot Frisius on the slopes of the Costa Rican rainforest. In the wild, small flocks of fiery-billed aracaris forage for fruit and insects in the humid forests. As many as five adults sleep together in old woodpecker holes—tails folded over their backs—and share parenting responsibilities. At Amigos de las Aves, those responsibilities fall to Richard and Margot.

The fiery-billed aracari *(Pteroglossus frantzii)* breeds only on the Pacific slopes of southern Costa Rica and western Panama.

Underfunded, Amigos de las Aves could not provide the high level of ongoing care Stravinsky required. So he moved to Lenny and Joan's finca. Lenny built a special cage with a high floor and an orthopedic perch. He situated it near the kitchen door, so the forlorn bird would have regular company and eye-level stimulation. Stravinsky eschewed his special perch, but gradually developed a relationship with Joan and Lenny.

Lenny introduces me to Stravinsky one morning at dawn, and I watch as the bird hops on his one good foot and balances on the stump that is his other leg. He doesn't screech like the larger birds, but chatters a quiet greeting. Lenny feeds Stravinsky a few slices of banana and fills his bowl with diced watermelon, pineapple, and papaya. Stravinsky eats the banana first. It's his favorite food. Sometimes, Lenny says, Stravinsky purrs when he eats it.

The next day, Joan opens the cage door and shows me how she makes a game with the first two fingers of her left hand, scissoring Stravinsky's beak. He stutters softly and nudges back, caressing her fingers with his beak. Then Stravinsky licks her fingers with his slim brown tongue. Joan says he likes the salt on her skin. She tosses a single piece of kitty kibble toward Stravinsky, and he catches it in midair. They play well together.

That evening, when all the other birds perch, feet and legs locked in sleep, Stravinsky slides his long, elegant bill through the bars of his cage and leans against it for support. That beak is a beauty: the upper mandible is vermillion, fading like a sunset to yellow-green, and then again to violet at the base. A crisp daffodil-colored band marks the line where bill meets face on Stravinsky's left side—his good side—but runs ragged and muddy on the right.

I wanted to find out more about Stravinsky, so Joan and Lenny took me to visit Amigos de las Aves, the eight-acre rescue center where the bird grew up. It began as the Flor de Mayo botanical preserve, founded by Sir Charles Lancaster, renowned naturalist and botanist to Queen Victoria. Sending exotic specimens home from the colonies was considered a noble scientific pursuit, and Lancaster procured thousands of plants from Central America for the Royal Botanic Gardens at Kew, which now boasts the world's largest collection of living plants.

Today, the center's history is commemorated by a single, modest wrought-iron sign that uses a few weak curlicues to decorate the rusting letters of the words "Flor de Mayo." The dark, damp grounds are populated with tall palms, giant climbing philodendrons, and uncountable epiphytes—plants that depend on others for their physical support. Orchids hide their beauty in the dripping undergrowth.

Richard and Margot transformed the silent botanical preserve into a lively avian rescue center, Amigos de las Aves. Years ago, Richard built special outdoor cages for the birds, using wire screening with squares of just the right size for the birds' feet to grasp comfortably.

The couple's reputation as animal rescuers and rehabilitators grew, and now Amigos de las Aves is inhabited by hundreds of "donated" birds—former pets who turned out to be too noisy, too smelly, or too aggressive for their owners' tastes. Parrots, macaws and other colorful exotic birds that were confiscated from poachers—presumably destined for the illegal pet trade—ended up here, too. Many of the birds at Amigos de las Aves can be re-released, if habitat is available. Some, like Stravinsky, cannot.

Amigos de las Aves specializes in the rescue and breeding of macaws. I saw at least two hundred on the grounds, and learned about the conservation challenges they face. One of the species, the great green macaw, is the second-largest parrot in the world; its wingspan is well over a meter. In the wild, great greens can live for more than fifty years. They usually fly in pairs or in small family groups.

The great green macaw is also called Buffon's macaw.

Millions of great green macaws once migrated between Honduras and Ecuador, their flight following the fruiting of the *almendro*, or wild almond tree. Today, ornithologists estimate there are at most 250 individuals left in the wild, and fewer than a third of those are breeding pairs. Amigos de las Aves is the only place that has been able to breed great greens in captivity. But successful captive breeding does not solve the problem. There is nowhere for the great green macaws to go.

Logging has destroyed most of their habitat. Seventy of Amigos de las Aves's great greens are ready for release now, awaiting appropriate sites. They need an adequate natural food supply, nesting sites, and the guaranteed safety of a private lodge or protected reserve. In the meantime, most live in small cages, too crowded to mate. Overcrowding makes them cranky. The longer the macaws are here, the more difficult it will be for them to find food in the wild, to build nests, reproduce, and avoid predation. The staff and volunteers at Amigos de las Aves are searching for release sites.

The visit left me discouraged, but Joan was philosophical. "The macaws are not destined to live on this Earth for very long."

Joan is right, I thought. *Every species ends. The only question is when, and how much it takes with it.*

The next morning, Lenny feeds Stravinsky his fruit. Stravinsky begins eating the banana, but when Lenny moves on too quickly to feed the other birds, Stravinsky stops eating and bangs the bars of his cage with his beak in protest. Lenny returns and rubs Stravinsky's outstretched bill with one finger, petting the top of his tiny head with another. They have developed a sweet ritual.

The days pass, and I continue to visit Stravinsky. I learn to stand a few feet back from the cage, so as not to frighten him. I speak quietly, tell him where I live, muse about the weather and upcoming elections. I admire the rhythmic pattern of shallow, saw-like serrations along the edge of his powerful bill; they would have been useful for feeding in the wild. In the evenings I stop by to bid Stravinsky goodnight, but he is usually already asleep.

One warm afternoon Stravinsky bathes in front of me, balanced on one foot, his short dark wings beating like miniature outboard motors, wet belly feathers wilting in a bedraggled mess. He uses his beak to splash water out of the shallow bowl in his cage. The spray is cool; Stravinsky is aiming straight at me. I feel like he's trying to get my attention.

After the bath, I watch Stravinsky hobble over to his fruit bowl. He stumbles once, then continues. The little bird picks up a cube of watermelon with the very tip of his bill. In one fluid motion, he opens his beak, tosses the fleshy fruit back into his mouth, and swallows. Then he picks up a second watermelon chunk, hops back over to where I stand watching, and extends his bill out through the black cage bars toward me. He is still holding the watermelon.

Is Stravinsky offering me his food—food that he cannot catch for himself, that he is dependent on Joan and Lenny to provide? If so, I am humbled by such trust and generosity. I am also perplexed: Lenny had asked me not to feed the birds, but he hadn't said what to do if a bird tried to feed me. Should I accept the soft, pink piece of fruit? I am not as trusting as

Stravinsky is. I back away from the gentle aracari and his powerful, serrated beak.

Stravinsky looks at me and waits, a large drop of liquid wobbling at the base of his bill. It looks like a teardrop. He balances on one leg with his good side facing me.

I look back and reconsider. The red band across Stravinsky's belly flashes like a ribbon of courage. His eyes shine clear and curious, with an inky pupil polka-dotting the center of each bright white iris. That droplet on his beak is not a tear; it is juice from the watermelon, a small, fragile offering. In this simple gesture, Stravinsky gives more than food: a reminder that even the tiniest life, scarred and imperfect, can still offer generosity, courage, and wonder.

Lenny walks over to the cage, sticks his finger in, and strokes the bird's back. Contented, Stravinsky closes his eyes and purrs.

An Australian farmer holding one of the
Giant Gippsland Earthworms in 1979.
This was a short one.

The Giant Worms of Gippsland

Twenty-foot earthworms, a third-grade dare, and the edge of extinction

In third grade, my classmate Jimmy claimed he would eat a live worm—for twenty-five cents. None of us knew where he got the idea, but we pooled our nickels one rainy spring day when worms had surfaced to escape the sodden soil. The sidewalk was littered with them, wriggling and ripe for the picking.

Soft, swollen, and the color of bruises, the worms were irresistible. Jimmy picked up an extra-fat one, flung his head back, and swallowed the creature whole. We debated for weeks whether swallowing counted as "eating," or if chewing should have been mandatory. Mostly, we wondered how it felt going down—could he feel it wriggling in his throat? Jimmy insisted he could.

That moment, and the ensuing debates, left a lasting imprint. Perhaps it was the worms' ubiquity, their wriggling vulnerability, or something about their shape that enchanted me. Whatever the reason, I was hooked on annelids, and the memory resurfaced decades later on the Bass Highway in southern Australia. There, a sign beckoned: "Wombat World—Live Wombat Display—Bass Giant Worm Museum." Naturally, I had to see the worms.

I didn't know at the time that reports showed these earthworms, *Megascolides australis*, are the largest in the world, reaching a length of more than twenty feet. I didn't know the giant worms were called *karmai* by the Aboriginal people, or that the nearby village of Korumburra held an annual Karmai Festival every spring for more than thirty years, at which the worm was celebrated with a lively carnival complete with an Earthworm Queen. The climax of the festivities was when local children hoisted a 100-foot "worm," resembling a Chinese New Year's dragon, onto their shoulders and paraded it through town.

I entered the museum, a one-of-a-kind attraction: a 300-foot-long building shaped like an earthworm. Inside, I discovered the wonders of the Giant Gippsland Earthworm (or GGE). They live underground in complex, permanent burrows— one worm per burrow—feeding on roots. Rarely, they poke their heads above ground to forage or escape a flood.

Remembering little Jimmy and the worms that

crowded sidewalks at school when it rained, I wondered what it would be like to stand in the GGEs' territory during a heavy rain. Would agitated twenty-foot worms cover the ground? Would they be searching for something—a tree or a fence post or a human leg, perhaps—to slither up in order to escape the flooding?

Inside the museum I crawled through a human-sized replica of the earthworm's burrow, designed to help people empathize with the life of a worm. The tunnel was narrow and dark, distinctly lacking in right angles, and seemed to go on forever. The air smelled stale. I worried about encountering organic matter. It felt like a lonely life.

There was much more to discover at the Bass Giant Worm Museum: Each worm has 300 to 400 body segments and is pinkish-gray except for the head and front third of its body, which is purple. The GGE's average size is five to ten feet in length and about an inch in diameter. But it is almost impossible to measure the worm's actual span accurately, because it can stretch to more than double its resting length. "Fresh body weight" is a more reliable indicator of size; adults average around seven ounces. For comparison, that's the same weight as a cup of sugar.

The museum displayed a few old photographs of people holding the worms. The most impressive showed three farmers standing in a field—each man two arms' lengths apart from the next, and all holding one specimen that was easily fifteen feet long. They were struggling to support the GGE, which drooped precariously.

Of course I wanted to touch one of these stupendous creatures, and there they were, eight live specimens housed in giant glass tanks filled with their preferred substrate, sticky blue-gray clay. One rested against the glass, head raised, purple and glistening. I marveled at the segmented body, the soft, supple skin, and the sheer improbability of survival for a creature so large, so slow, so geographically isolated.

But the museum's live worms were not available for touching. The creatures are so fragile their skins "can burst as they try to escape capture, and even slight bruising may result in death." I tried to imagine the skin of a twenty-foot worm bursting—how would that happen? Would it stretch so much in its escape attempt that the skin thinned and burst in one spot? Or would some stress-induced chemical reaction occur that made the whole creature explode? Would it all happen in an instant, or would the worm die a slow death? The museum offered no details.

I did learn that *M. australis* prefer to live under stream banks and on south- or west-facing hills and require consistently wet ground in order to move and breathe in their underground mazes. In fact, if you stomp the ground above them—or bang the ground with a spade, which is the method recommended by two earthworm scientists named Van Praagh and Hinkley in their extensive 2002 report about the natural history of *M. australis*—you will hear a loud gurgling noise coming from beneath you. This is the sound made by the worms moving through their wet tunnels to escape the disturbance you have created. The museum played a recording of the sound, which is similar to the gurgling slurp made by the last few ounces of bathwater flowing out of a tub. It was on a loop, and quickly became annoying.

The Victorian Flora and Fauna Guarantee Act of 1988 lists the GGE as a threatened species, and the government is actively working to save it. I found the National Recovery Plan for the Giant Gippsland Earthworm, a 31-page overview from the Victorian Department of Sustainability and Environment. It details the worms' natural history, management practices, and a five-year Recovery Action Implementation Timetable estimated to cost AUD $1.675 million.

M. australis is hermaphroditic, possessing both male and female organs. Even so, two worms are needed for fertilization, and a pair produces at most one egg per year. Incubation takes another year, and the baby worms measure a remarkable eight inches at hatching. It takes them up to five years to reach sexual maturity

—meaning the species reproduces very slowly.

As with many of Australia's native flora and fauna, European colonization meant the decline of the GGE: Reports from early Gippsland settlers include accounts of "ploughed fields red with blood," describing the carnage wreaked upon the worms. There are photos to prove it, too—horrifying shots of muddy fields carpeted with the bloody remains of chopped-up worms.

As I take this all in, it occurs to me that I'm witnessing a fleeting moment in the history of the earth—that moment when a particular species teeters on the edge of extinction. These creatures probably will not make it to the next century. They live only here, in a small area of just 170 square miles in Victoria, within which they are only patchily distributed. This kind of geographic isolation means there's little opportunity for genetic exchange, but it's difficult to know, because there is no method for monitoring *M. australis* without injuring or killing individual worms.

In addition to cultivation and other soil disturbances, drought, use of pesticides and fertilizers, and even the collection of individual worms threaten the GGE. (Computer simulations suggest that the death or removal of just three earthworms from a local population would cause the extinction of that population within fifty years.)

The captive worms weren't safe either. I later

learned that the Bass Giant Worm Museum has been permanently closed, citing animal welfare concerns. I wonder what became of the worms—were they released into the wild? If so, were their captors able to find a suitable place in which to relocate them, complete with a year-round stream and blue-gray clay and a south- or west-facing hill? Were there any territorial disputes when the new worms were introduced? Were the worms able to adapt to their new environment, which was undoubtedly less hospitable than the temperature- and humidity-controlled museum tanks? Are they reproducing, and maybe even evolving into worms that can survive in, say, the red clay of a north-facing hill? I'll never know.

As for my third-grade friend—well, I did hear what became of him. Jimmy grew up, went off to college, and earned a Ph.D. in molecular biology. These days he teaches and conducts research at a prestigious university. I have no idea what he eats for lunch.

Baby Rastus highjacked my heart, and Janet set my head spinning.

Domesticating Janet

On the ethics of riding a two-ton orphan

"Put your hand in the elephant's mouth," Daniel suggests. Rastus, a five-month-old orphan, barely reaches my waist—but he already weighs 340 pounds. Big enough to knock me over. Big enough to crush me.

I'm visiting a small compound in Zimbabwe where orphaned calves are fed, sheltered, bathed, and given a home. Five-month-old Rastus is one of them. So is six-year-old Janet, whom I'm about to meet. Daniel is one of the *mahouts*, or elephant handlers, who work at the compound. He's in his twenties and wears park-service-green shorts and a khaki short-sleeved shirt. A green baseball cap protects his eyes from the blazing African sun.

I stall, unsure I want to risk my hand. "His ears are amazing!" I exclaim instead. One spreads like a cabbage leaf—smooth, veined, warm, and velvety. Everywhere else, Rastus is soft and a little hairy.

"Elephants use their ears to regulate temperature. That's why they're so large—to provide surface area."

I peer at Rastus' closed mouth. "Does he have teeth yet?"

"Not yet. He's still teething; his first set of molars will break through in the next month or so."

I remember my son teething—the sharp ridges, the drooling, the relentless gnawing. Elephant-scale teething, I decide, is a level of discomfort I'd rather avoid.

"We found Rastus abandoned in the jungle," Daniel continues, looking down at his charge. "His mother must have been dead. If we hadn't brought him here, Rastus would've been eaten by a lion or a hyena, or he'd have starved." He pats Rastus' head. "All the elephants here owe their lives to rescue."

Rastus lies down on the sparse crabgrass lawn, rolls onto his side, and rests his head in Daniel's lap, wrapping his trunk around the mahout's neck. He's a baby, but he knows how to demand affection. I sit down beside them, amazed at the bond. "He's very attached to you," I say.

"Yes, elephants are social animals. A baby growing up in the wild would be within touching distance of his mother all the time."

"How do they interact in captivity?"

"The adult females here love Rastus, especially the young ones. They fawn over him like teenaged girls playing with a cute little baby. But he's bonded to humans because we feed him. He needs to be with someone 24/7."

"Really? What about at night?"

"Rastus is a typical baby; he won't sleep alone."

"How long do they sleep at night?"

"They lie down for about two hours, and you won't believe how loud they are."

I try to imagine the elephants' sleeping sounds—maybe an occasional grunt or sigh, their bodies heaving as they roll over or rearrange a leg, the hay beneath them scratching and rustling. But that's not the kind of noise Daniel is referring to.

"Elephants snore very loudly!" he explains, laughing. "The adults sleep over there in the stable, where they can touch and talk to each other through bars. But they can't get to each other's food—we don't want fights. During the day they might doze resting against a tree, to relieve some of the weight of their tusks."

"How big do their tusks get?" I glance around the compound. A dozen or so elephants mill about in a small stable. An equal number of mahouts, all dressed like Daniel, watch the elephants closely.

"Years ago, an average elephant tusk weighed about a hundred pounds—per side!"

"That's a lot of weight to carry around. But these don't look all that big."

"Well, the tusk extends up into the head, so only half of it shows. Also, hunters have killed the animals with the largest tusks, eliminating them from the gene pool. Now the average tusk weighs only about forty pounds. The elephants are losing the tools they need to eat bark."

I'm alarmed to realize we humans have not only encroached on elephants' habitat, not only killed individuals for their ivory tusks, but have also actually influenced their evolutionary path, effectively selecting the smaller-tusked animals for reproduction. I have lots more questions, but Rastus snuggles up next to me and nuzzles me with his trunk. Like a human baby, he needs attention—and demands it.

Caring for these orphans is clearly a big job. "Just feeding him must take a long time," I observe. Rastus drinks milk every few minutes from the liter bottle Daniel offers him. It's topped with a thumb-sized nipple, and Rastus is a good sucker. "How much does he drink?"

"Forty to fifty liters of milk—every day."

"Of elephant milk?"

"No; we don't have access to that. We feed Rastus

a mixture of nonfat cows' milk and desiccated coconut, with added calcium and multivitamins. It's easy to digest and provides plenty of calories. We had to do a lot of experimentation to get the formula right."

"But couldn't you use the recipe from a zoo somewhere?"

"Not really. The success rate—all around the world —for raising baby elephants without their mothers is very low. We're doing surprisingly well here."

I'm trying to take notes, but Rastus will have none of it. He reaches over and fondles my arm with his trunk—snuggling and nuzzling, rubbing and bumping his head against me. I'm smitten.

Daniel can see that the elephant is comfortable with me. "Go ahead; put your hand in his mouth."

I screw up my courage and slide my right hand, palm down, into Rastus' open mouth. That's easier than it sounds, because even a baby elephant's mouth is quite large. It's warm and soft and almost dry inside, not wet and slobbery like I'd expected. Rastus chomps down and gums my hand—just like a human baby that's teething, only not as rough. It's as though the elephant knows he has to be gentle with a human being.

Rastus continues gumming my hand, obviously enjoying himself.

"It's almost time for the elephants to go for their

walk," Daniel says, standing abruptly. "Do you want to ride one?"

I am not sure about this for a number of reasons, but my hesitation is mostly about my own safety, not the impact on Janet. I don't yet fully appreciate the ethical implications of elephant rides. "Is it hard?" I ask.

"No, it's easy. And the elephants don't mind. Robert is a mahout; he will take you; I'll stay with Rastus." I wonder how Daniel knows whether or not the elephants mind. He probably sees the question in my expression. "Don't worry. We train them very gently—not like Indian elephants."

I hadn't realized Asian and African elephants were trained differently, and consider whether it might have to do with the elephants' temperaments or with the mahouts' differing traditions. I don't worry much about it at this point, however, because something else is on my mind. I don't want to leave Rastus. *He's my baby!* I can't leave him.

But Rastus is back to gumming Daniel's hand. He's Daniel's baby, not mine. My heart drops as I turn toward Robert, and we walk over to the elephant I will ride.

"This is Janet," Robert says, introducing us.

Janet has gigantic ears—one of them would easily wrap around my body like a blanket. My head comes up to her yellowing tusks, which are brighter at their points than at the base where they intersect with her thick skin. The left tusk is half a yard long; the right one is shorter—closer to a foot. Their ends taper to blunt points just inches from my face. Janet's own face is crisscrossed with a vaguely geometric pattern of wrinkles. A dark smear of dampness escapes from the inner corners of her tiny eyes. Her expression is impenetrable.

"She's kind of small ..." I venture.

"Janet's six," Robert says. "Get on. It's easy."

I step over grassy clods of elephant dung, climb a tall, sturdy mounting platform, and swing my leg over Janet's back. She wears a leather saddle, like a horse's, only elephant-sized. Robert scrambles up after me but sits in front of the saddle, just behind Janet's ears.

We sway along in a slow caravan with a dozen other elephants, all about the same size, each with a passenger and a mahout. A half-mile loop leads us through the veld to a small lake and then back again. Janet and the other elephants stop frequently to browse on clumps of grass or leafy branches. At the lake they mill around the muddy bank, slurping drinks among the reeds and lily pads. The mahouts don't pressure or hurry them, and the elephants seem comfortable. Actually, they seem much more amenable to carrying passengers than many horses I've known.

"Why is Janet's left tusk longer than her right one?"

"Are you right-handed, or left-handed?"

"What?"

"Elephants are like humans; they are either right-tusked or left-tusked. The tusks get a lot of use—digging up trees and stripping branches—and the one that's used most gets worn down. Fortunately, they keep growing until the elephant is about thirty-five years old."

"What's an elephant's life span, anyway?"

"It depends on their teeth," Robert says. "They have six sets of molars. When the last set wears down, maybe in their forties or fifties, they starve."

I'd never thought about the end of life for an elephant. I look down at Janet, wondering about the quality of her life so far. Some people say it's cruel for humans to ride elephants, and I hope I'm not rationalizing in thinking it's similar to riding a horse. That doesn't seem cruel. Is the difference that horses are domesticated, but elephants are not? In the long run, that doesn't seem like much of an issue; the process takes generations. After all, horses were wild once, too.

"Do they mind giving us rides?" I ask Robert.

"They give visitors a ride, and visitors pay for their upkeep. The elephants work for a living, just like we do." It's a tidy explanation, and part of me wants to accept it. Another part can't quite forget that only one of us in this arrangement had the option to walk away.

These elephants are tame; they've been trained to accept the presence of humans, and even to let us ride them, but that isn't domestication—not yet. Domestication takes generations and changes the animals themselves, not just their behavior. This might be a first step, though. Have I participated in initiating that process? If so, is it ethical? The question lingers.

And it raises another, broader, consideration: The process of domestication may seem unnatural, but what if the animal *itself* initiates it—does that change anything? Evolutionary biologists believe that when dogs began to diverge from wolves, some 15,000 years ago, it was a process initiated by the wolves, not by humans. Wolves that came closest to human encampments got more food, which encouraged them to spend even more time near humans, which got them more food, which meant stronger wolves that became better breeders—and so it went.

Besides giving me a lot to think about, the experience with Janet and Rastus deepened my admiration for elephants. They play, cooperate, teach, grieve, and even seem capable of broken-heart syndrome. Humans have used them as beasts of burden, war machines, and trophies. We've destroyed their homes and reshaped their genetics—and we're in the process of decimating the very planet they live on.

Rescue or not, what do elephants gain—or lose—when we draw them closer to our world? Janet and Rastus remind me that our love for wild things is never simple, and that every touch—every ride, every bottle, every hand in a baby's mouth—leaves a mark. Somewhere, an orphan calf is leaning into a mahout's hand, and a small gray trunk is reaching for a human it trusts. Whatever else we decide about ethics and evolution, that fragile, complicated bond is now part of the story too.

Being the fastest may not be enough to save these threatened felines.

Mark of the Cheetah

Earth's fastest land animals
and their unlikely allies

The sun is harsh, and desert grit dusts my lips. A pale expanse of tawny sand, broken by clumps of vegetation, stretches toward towering cliffs on the horizon. Suddenly, not fifty feet away, a cheetah lifts from the ground. He appears as if from the earth itself and then he's gone—streaking past in a blur, accelerating from zero to sixty in three seconds. My heart jolts with him; he's past the edge of my vision long before I remember to exhale.

The fastest land animals on Earth, cheetahs can sprint at more than seventy miles an hour; they're built unlike any other big cat: all spine and suspension, loose-limbed and elastic, with "floating" shoulders and a backbone that coils and unspools like a spring with each stride. Every part of this feline—from limbs to lungs, nose to tail—is optimized for speed. Their claws, not fully retractable, function like the cleats on running shoes.

I had been fascinated by cheetahs since childhood, studying their long, lean lines in the pages of library books. It wasn't just the fact of their world-class speed that transfixed me. Cheetahs are unique in several other ways: First, they cannot roar. Instead, cheetahs purr, chirp, and bark—sounds that seem oddly intimate for an animal capable of vanishing in seconds.

Like housecats, cheetahs purr for several reasons. Typically, it signifies contentment, relaxation, and bonding—but purring can also be self-soothing. And it's been suggested that the low-frequency sound may help with muscle and bone healing after intense sprints.

They project graceful confidence, walking with the assurance of an insolent princess. And they sport a distinctive black teardrop mark beneath each eye. Besides embodying a cheetah's too-cool attitude, the black streaks reduce glaring sunlight, so cheetahs can hunt during the day, thus avoiding competition from powerful nighttime predators such as lions, hyenas, and leopards.

Despite all these elegant adaptations, cheetahs are among Africa's most endangered big cats, with only about 7,000 remaining in the wild. Most of them live in Namibia—a leader in conservation and the first country to include protection of the environment in its constitution. It's known as the cheetah capital of the world.

I was traveling there to learn more about best practices in conservation, and arranged a visit to the Cheetah Conservation Fund's (CCF) research and education center near Otjiwarongo, roughly 150 miles north of Windhoek. It's one of the best places in the world to learn about cheetahs, observe them, and talk with those dedicated to protecting them.

Founder and Director Dr. Laurie Marker fell in love with cheetahs while pursuing graduate research in veterinary medicine. She established the Cheetah Conservation Fund, in 1990, and by 1994 had purchased a farm to create a place where cheetahs could live in balance with people and the ecosystem. Laurie Marker is often described as the Jane Goodall of cheetahs—devoting her life to understanding these animals and ensuring they can continue to live in the wild.

I met Dr. Marker, a commanding figure with big hair and an even bigger smile, at CCF. She radiated the calm focus of someone who has told this story a thousand times and still believes each telling matters. Dr. Marker generously took time to answer my questions and show me around, explaining the organization's many projects as we walked. The cheetahs were our first stop. The one I had glimpsed earlier was exercising—chasing a lure, like those used in greyhound racing, around a half-acre paddock.

Along with twenty or so others, he lives at the center as an "Ambassador Cheetah."

"Most were orphaned before they were old enough to learn to hunt," Dr. Marker said. "One had been kept as a pet for five years. None of them would be able to survive if they had to fend for themselves."

These ambassador cheetahs are the public-facing heart of CCF's research and education center. Lazing in the scant shade of a thorn tree after mealtime, t hey are surprisingly well camouflaged. The cheetahs' tawny fur echoes the colors of ground and dried grasses, and their dark spots mimic dappled sunlight. A tail flicks. An ear turns. I find myself scanning the landscape for movement, realizing how easily a hundred-pound cat can disappear—not because it is hiding, but because it belongs.

Understanding a wild animal's subtle signals is a crucial skill here at CCF, because the staff sometimes work in close proximity to the cheetahs. Years of careful observation—ear angle, tail position, pupil size—have taught them exactly how near they can safely be. I flinched the first time a bare hand moved toward a spotted head. For a cheetah that needed eye drops, a handler reached through a wire fence with tongs full of food while another leaned in swiftly to administer the medicine. One afternoon a group of handlers drove a pickup into a huge fenced-in paddock and tossed meat out for the cheetahs that lived there. These cheetahs could easily have jumped into the back of the pickup demanding more food—but they didn't.

The cheetah orphanage/education center is only a small part of CCF. Most cheetahs, of course, live in the wild, so that's where most of the effort needs to go. "Here in Namibia," Dr. Marker said, "about eighty percent of cheetahs are outside protected areas, primarily because their native ranges are so large. Our protected areas aren't big enough to sustain a population of cheetahs within them. They're on lands where there are lions, hyenas, and leopards, and there are people and livestock."

"So how do you have any control over what happens on the land?"

"We teach rangeland management, because it's crucial to have enough grass. Cheetahs live in some of the driest landscapes in the world, and where you have these arid landscapes you also have very poor people."

Dr. Marker had figured out early on that a big part of cheetah conservation involves human interaction. When she started CCF thirty years ago, the biggest danger for cheetahs in Namibia was farmers. "At that time," she said, "they were shooting 800 to 900 cheetahs each year, because they thought cheetahs were killing their livestock."

Dr. Marker suspected the threat was more perceived than actual. She needed to find out whether cheetahs really *were* killing large numbers of livestock. So CCF developed a program to scent-train dogs so they could help humans find cheetah scat, just like dogs can detect drugs, bombs, and diseases. I met one of them, a seven-year-old Belgian Malinois named Camina.

Lean and alert, she looked a lot like a German shepherd, but with a tawnier coat. Camina moved with precision. Her sharp ears twitched; her eyes glowed amber in the sunlight. My fingers itched to scratch behind her ears, but because Camina is a working dog, she barely spared me a glance.

Conservation scientists at CCF worked with Camina, and dogs like her, to locate cheetah scat in various areas, then tested the scat to find out what the big cats had really been eating. I'd expected the lab work to confirm the farmers' assumptions—sleek cats slipping into corrals at night. But the data told a different story: The cheetahs were mostly eating wildlife. The problem wasn't predation; it was perception.

Simple reassurance about the cheetahs' preferred diets was not enough to calm the farmers, though, and multiple steps would be required to convince them to stop shooting and trapping the big cats.

"The people who live here are very rural or pastoral, and they have goats that they will carry with them, for eating, primarily—meat on the hoof. So we also teach livestock management—having a herd or having calving seasons, corrals, and proper healthcare of these animals," Dr. Marker explains.

The CCF began to work with farmers on regular vaccinations and de-worming, wound care, managing breeding, and sheltering young, sick, and injured livestock. Healthier herds are less attractive to predators. "This is a global issue. We all think that predators are wanting to eat our livestock, but actually, if there's predator problem, it's probably related to your management, and not that a predator just wants to eat your livestock. They would much rather have wildlife."

Dr. Marker began another canine-based initiative, CCF's now-acclaimed Guardian Dog Program, in 1994: They train Anatolian shepherds to protect sheep and goat flocks. These dogs are big—they weigh up to 180 pounds—and their bite, at 734 pounds per square inch, is even more powerful than that of a lion. "We selected this breed because it lives in vast, open areas up on the Anatolian Plateau. They are independent-thinking dogs, that don't need a herder or a person to tell them what to do. They've been doing this for 5,000 years, and they're incredible at it."

"They mark their territories so predators know where they're living, and avoid them." If one does venture near the herd, "the dogs position themselves between the herd and the potential predator, and that's pretty much that. No one wants to have to go through an Anatolian shepherd to get to dinner. We call it *Dogs Saving Cats*."

CCF trains and supports these dogs from puppyhood, providing veterinary care, monitoring bonding with herds, and ensuring they are not malnourished or mistreated. Over 800 dogs have been placed to date, with demand still high.

All this work has paid off: Livestock losses are dramatically lower, and farmers no longer feel the need to exterminate cheetahs. CCF has developed a scalable, humane, systems-based approach to conservation—protecting predators while safeguarding people's livelihoods—that can be replicated in other areas with other threatened species.

On my last afternoon at CCF I sit on a low wall, watching two cheetahs that live in the paddock just outside my cabin. Siblings Hamish and Kwili are about twenty feet away, easy to observe as they lie, almost motionless, on the sparse grass. The sun is low, shadows are lengthening. I talk quietly to the cats, telling them about cats I have lived with, asking about their own lives before the CCF, wondering about their future. Sitting in the shifting Namibian light, I'm struck by the elegant simplicity of it: Sheep graze. Dogs guard. Farmers watch. Beyond the fence line, cheetahs run. The balance is not perfect or permanent —but for now, it holds.

One of the cheetahs rises and walks toward me. Slowly, deliberately. It ambles all the way to the chain-link fence that surrounds the paddock, separating visitors from the cheetahs, stops directly in front of me, and sits. It begins to purr. The sound is soft, higher-pitched than you'd expect, and entirely disarming. I feel a little giddy as my face splits into an involuntary grin. And then, something remarkable happens.

The cheetah rubs its forehead up against the fence right at my eye level. It's a hard rub, like a familiar pet demanding affection. The force of it makes the fence wobble and smash into my face. I hadn't realized I was so close! Startled, I jump and retreat, barely registering that my glasses have fallen to the ground, that I probably felt the cheetah's fur, and that I have made at least one poor decision in the last few minutes.

The next morning I wake up to a *perfect*, teardrop-shaped bruise near the inner corner of my right eye. My glasses must have bumped against my cheek, the plastic nosepad creating the teardrop. It looks like the tattoos some people are marked with in prison, but I choose to think of it as the Mark of the Cheetah: my own initiation into the wild. The bruise disappears in a few days—unlike my memory of this indelible connection, which has marked me forever.

:::::

Witnessing the ways humans and animals can work together to survive left me hopeful—but I was about to meet an animal for whom the scars of human conflict were permanent...

Sama is the saddest elephant I have ever seen.
I couldn't look at her without crying.

Lucky Sama

A meditation on the life of a three-legged elephant

Sama is the saddest elephant I have ever seen. A small adult, she has only three usable legs and leans desolately against a sturdy metal fence at the Pinnawala Elephant Orphanage in central Sri Lanka. A hundred or so tourists crowd around the dusty corral that houses Sama and three adorable baby elephants. Most of them are admiring the babies, and a few lucky visitors even get to feed the youngsters from giant baby bottles. But I am looking at Sama.

Sama's right front foot and about six inches of her lower leg were blown off when she stepped on a land mine during the Sri Lankan civil war. Although the injuries themselves have long since healed, the elephant's left front leg is severely bowed from carrying her poorly distributed weight, and her back is grotesquely twisted for the same reason. But it is

Sama's eyes that affect me the most.

She gazes at me with the saddest expression I have ever seen. I am surprised to discover that an elephant can be expressive, but there she is, four feet away, looking at me with a mixture of resignation and anguish. She must be in a great deal of pain. We stare at each other for a minute or so until I can't bear it any longer and turn away. I know it sounds silly, but I don't want Sama to see me cry.

I also don't want her to see the pity in my gaze. I've read a lot about the emotional life of elephants, and there's no doubt among animal behaviorists that elephants exhibit empathy. They respond to each other's pain or physical problems; they assist one another, comfort one another, and grieve for their dead. Standing in front of Sama, I find it impossible to dismiss those findings as mere anecdotes.

I don't want the nearby humans to see me crying, either, so I move on to explore more of the twenty-five-acre Pinnawala Elephant Orphanage. It's on an old coconut plantation in the forested hills of central Sri Lanka, between the ancient royal capital of Kandy and the present capital, Colombo. And it is home to the world's largest herd of captive elephants.

Sri Lanka, the 25,000-square-mile tropical island just off India's southeast shore, was once a paradise for Asian elephants. Before the British invaded in

1815, an estimated 30,000 of the pachyderms lived on the island. But according to the Sri Lanka National Zoo, in the 1960s, "following nearly a century of game hunting and jolly slaughter by the British colonialists," the elephant population dwindled to the point of near extinction. This tragedy prompted the Sri Lankan government to found the Pinnawala Elephant Orphanage.

British colonialists were not the only threat to elephants in Sri Lanka. Drought can be a serious problem there, so over the centuries various rulers have built tens of thousands of reservoirs, called "tanks," across the country. Elephants—especially small ones—regularly fall into the tanks and get stuck. Add to that the problems of poaching, habitat loss and fragmentation, construction of roads and railways, and human encroachment to the forest for settlement and agriculture, and the situation looks grim.

Life at the Pinnawala Elephant Orphanage, on the other hand, seems quite agreeable. As much as possible, the daily routine simulates conditions in the wild. The elephants are allowed to roam freely during the day, and have formed a herd structure with stable social interrelationships. Even though each adult requires about five hundred pounds of food each day, there's plenty of jackfruit, banana, coconut, tamarind, grass, foliage, and other vegetation for them on the reserve. There's plenty of fresh water, too, which is a good thing, because elephants drink up to fifty gallons a day—about as much as a standard bathtub holds.

Bathing is a primal pachyderm pleasure, and twice a day—from ten o'clock to noon and again from two to four—*mahouts* (keepers) walk the elephants a quarter of a mile through the village of Pinnawala and down to the wide Maha Oya River, where the elephants cool off, play, or just relax. It's one of their favorite activities.

One of the tourists' favorite activities is watching the bathing elephants. The individual animals interact with each other, splashing and squirting, sometimes lying down in the shallow rapids, rolling and lolling and generally looking like they're having a good time. From the riverbank, it's easy to believe that this is what elephant happiness looks like.

A natural environment, good care, and freedom of movement have afforded the elephants at Pinnawala opportunities to mate, and in 1984, the first baby elephant of Pinnawala was born. Today a scientifically designed captive breeding program is in place, and some of the elephants are third-generation residents.

Sama is not the only notable elephant at the Pinnawala orphanage. The oldest tusker here is Raja. Closing in on seventy years, he is blind in both eyes, weak, and at the end of his natural life. He is a silent monolith, not interacting with much in his environment except his mahout, whom Raja reaches out to touch with his trunk every minute or so. Perhaps it reassures him. Raja came to the orphanage as an adult, suffering from gunshot wounds he sustained in the wild. His tusks are huge, and part of one is missing. The mahout explained that the exceedingly long tusk was making it difficult for Raja to eat, so his keepers sawed it off. Like Sama, Raja carries the scars of human conflict in his body.

YouTube activists have raised concerns about the treatment of animals at Pinnawala, posting videos showing that some of the larger elephants' ankle chains have chafed or injured the animals. I didn't observe that problem when I was there; perhaps the publicity prompted a change.

If the Pinnawala Orphanage is a haven for elephants, a tiny shop in nearby Pinnawala village is a haven for tourists with a sense of humor—it sells specialty items made from elephant-dung paper: booklets, boxes, picture frames, journals.

The shop walls are adorned with handmade posters showing various stages of the papermaking process: The dung is already ninety percent fiber when it exits the elephant. It is boiled for three days—and let me just say I'm glad I wasn't around for that demonstration—colored for a week, blended for three to four hours, and spread across a small screen to dry in the sun for a day. It is then pressed into a much finer product, scented with cinnamon or lemongrass, and proudly packaged as elephant-dung stationery and elephant-dung notecards. I'm impressed with how people here manage their recycling needs and bring in tourism money at the same time. It is, in its own quirky way, a tiny act of turning waste into something useful.

I wanted to know more about Sama, so I did a little research after I got home. It turns out this elephant, whose name means "Peace," is known around the world. But much of the information available about her online is contradictory. Sama may have had the accident in 1996, when she was six years old. Or it happened when she was two. She may have arrived at Pinnawala in 1995, or perhaps it was in 1992 or 1984.

Information published in 2011, possibly from a 1998 research report, states that Sama "is now twelve and will suffer from considerable discomfort in the future due to changes in her spine, because of her unnatural body position, trying to balance the body weight on three legs."

Sama's situation was made worse by the fact that she was beginning to experience circulatory problems, as evidenced by changes in her pigmentation. Her keepers were also concerned about the possibility that Sama might develop psychological difficulties due to her medical condition, so they contracted with a German engineering firm to fit Sama with a prosthetic leg.

The Germans were excited about the undertaking. They set up a website for what they dubbed the "Lucky Sama Project," raised money and, in 2003, constructed a cylindrical trans-tibial (below-the-knee) prosthesis of steel and cotton, customized for Sama so that it fit perfectly.

Sama would have none of it; she managed to remove the contraption from her leg, breaking it in the process. The next day her doctor repaired the prosthesis and tried again, but with the same results. After a third try it was determined that Sama would have the final say with respect to the prosthesis, and that say was, *"No!"* Was her refusal stubbornness, or a declaration of what her own body can and cannot accept?

And what does the future hold for Sama and Raja? They'll live at Pinnawala for the rest of their lives. I can't imagine that either will survive much longer, but I'm glad to know they are as comfortable as possible, given their circumstances.

Some of the other rescued elephants will eventually be given to zoos. Many more will be "honored" by being presented to Buddhist temples, where they will be given starring roles in traditional religious processions. But even these conditions are exploitive: The animals must be taught to behave obediently in parades, submitting to human commands and control. They must walk for miles in stultifying heat and humidity, carrying their lavish—and heavy—

caparisons. They must endure drums, shouting, high-decibel fireworks, and other assaults on their sensitive hearing.

Many succumb to loneliness and depression. Elephants are intelligent, emotional creatures that have evolved to live in herds and need companionship. News reports of zoo elephants dying from loneliness have documented incidents in Japan, the Philippines, Spain, the United States, and elsewhere around the world.

As I leave Pinnawala, I ponder the future of the elephants. We are learning more and more about their emotional capacity, and some places—Pinnawala is an example—seem to have figured out how to treat them ethically. Online activists don't hesitate to step in with photos, videos, and pleas for action when they see mistreatment. Circuses are closing—due in part to citizen protests about inhumane treatment of elephants—and relocating their animals to wildlife refuges. I have some hope that captive elephants' lives will continue to improve.

But I doubt that there is enough time to save the species. At the turn of the twentieth century there were an estimated 100,000 Asian elephants in the world. Today there are fewer than 40,000, and their numbers are decreasing, mostly because of habitat loss. They are officially classified as Endangered. I think of Sama, "Peace," standing in her dusty corral. She has adapted to us, at great cost. Whether we are adapting in return is less certain.

Chaik, the one-thousand-pound brown bear I visited
just outside Sitka, Alaska

Prayer Bear

In the lair of Sitka's big brown bears

Chaik, a one-thousand-pound brown bear, lumbers toward me, stops, and rises to his full height—about ten feet. He looks directly at the man standing next to me and roars.

"Do you want to feed him?" the man asks.

We are forty feet away from Chaik, which would be a far-less-than-safe distance except for the fact that the hulking bruin is safely contained in a gigantic cement pit—more than half the size of a football field. It's one of two such structures, which are all that remain of an abandoned pulp mill just outside Sitka, Alaska. I am safely situated on the walkway at the top of the seventeen-foot-high wall surrounding the old pulp basin that's now Chaik's home. The bear is not the one who needs protection—at least not now. But that wasn't always the case.

I am visiting the Fortress of the Bear, and its founder and executive director, Les Kinnear, is showing me around. Les has been a licensed wildlife guide for more than two decades and knows more about bear behavior and biology than anyone I've ever met. He moves slowly, speaks quietly, and seems as comfortable in the bears' presence as he is in mine.

Six-year-old Chaik and his little brother Killisnoo, who weighs a mere seven hundred pounds, came to this nonprofit education and rescue center as cubs, shortly after it opened in 2007. Les got the idea for it after reading a newspaper article about the state having killed three orphaned bear cubs, who would have slowly starved to death without their mother. Shooting was—and still is—the only official protocol Alaska has for dealing with orphaned cubs, but Les thought there must be a better way.

He located two huge water-settlement tanks left from an old pulp mill just outside Sitka, abandoned in 1993. With their high walls, concrete floors, and controllable drainage system, they made perfect bear enclosures. After spending several years navigating a complex and often frustrating permitting process, Les established Fortress of the Bear. Local environmentalists protested the plan at first, but have since been satisfied that the bears here are well cared for.

"How much space does a bear need, anyway?" I ask.

"International zoo standards mandate at least 4,200 square feet for two adult bears." Les has the data on the tip of his tongue. "This space houses Chaik and Killisnoo in a 35,000-square-foot enclosure—that's more than eight times as large as is required."

A second enclosure that's 24,000 square feet is home to three more residents: brown bear triplets Balloo, Lucky, and Toby. A year and a half old at the time, they were not yet mature enough to survive on their own—and their mother had died from eating plastic bags she found in a garbage can.

Because they are circular, the pulp pits provide a "continuous environment." There are no unnatural corners to hem the bears in. The pits have dirt floors with small berms, low-growing vegetation, scattered branches, and the occasional tree stump. From above, it looks pretty much the same as the surrounding environment, except that the pits also include good-sized swimming holes, a bright yellow soccer ball, and a couple of old tires the bears like to play with.

"But wouldn't they be better off in the wild?" I persist.

"All the bears here were orphaned because of human behavior, and would've been euthanized if they hadn't found a home." It turns out that being orphaned because of human behavior is a common occurrence for bears. Their mothers might be hit by cars or shot by hunters. If a hungry mother bear is hanging around too close to humans and considered a nuisance, she might be destroyed for the sake of human safety.

"Is it even possible for bears and humans to live in the same area?" I wonder.

"*Yes!*" Les is emphatic. "Bears and people can live harmoniously in the same area; they have been doing so successfully on some Alaskan islands for decades. But it doesn't work so well," he warns, "when we entice bears with outdoor barbecues and pet food, and then shoot them when they come around foraging for table scraps." The cubs left behind are lost and terrified—traumatized without their mother. Because they're so anxious, it's difficult to place them in commercial zoos. That is how all the bears here arrived: not as curiosities, but as survivors.

I like the way Les answers. He may know more about bears than anyone else in the world, but he plays the information out slowly, like a fly fisherman. He never provides too much for me to absorb; in fact, he holds back a little. That just makes me more curious about the bears.

"What do they do here all day?" I ask Les.

He watches them for a few moments before answering, his face softening. I've seen this look before—it's that of a proud father. "The bears here are displayed in a way that reflects their innate behaviors

—they are not bored or pacing. You can see them interacting with each other and their environment."

The bear I'm looking at now, the small male named Balloo, appears to be interacting with himself. Les is watching me watch the bear. "Is that one, um ...?" I trail off.

"Yes, he's ... doing what boys do," Les says, and leaves it at that.

"What else do they do?"

"Killisnoo and Chaik search for food, explore, and play with the toys we provide. They also spend a lot of time in the water; they're more closely related to polar bears than to other brown bears."

DNA analysis suggests that the brown bear *(Ursus arctos)*, a species that also includes grizzlies, was an ancestor to polar bears, which then evolved to develop specializations—such as a camouflaging white coat and teeth suited for eating seals—for inhabiting the harsh Arctic. Here in Sitka, that ancient relationship lives on in the way Chaik and Killisnoo slip into the water as if it were home.

Killisnoo and Chaik are roughhousing, but Les assures me it's play fighting, nothing serious. They tussle in the water, wrestling and swatting at each other, splashing and growling for a few minutes. Soon they settle into what looks like comfortable companionship.

"How do you know whether they're healthy?"

"A veterinarian checks them regularly for large scars or tattered ears, which would be signs of serious fighting. He also checks their paws for cuts or other injuries, like broken nails or toes."

"Are the bears happy?" I want to know.

"Everyone asks that question, and no one really knows the answer." Les is hesitant to anthropomorphize. "What we do know is that these bears are well fed and secure. We've eliminated the stress of finding food. There's no need for them to defend their territory, to travel through other bears' territories, or to encounter larger bears that could harm them."

Even though they can't duplicate the demanding and often dangerous ways bears must secure food in the wild, trainers at the Fortress of the Bear can still encourage the bears to interact with their environment and hunt for their dinner. Rather than feeding the bears on a regular schedule, trainers hide food throughout the enclosures.

"What do they eat?"

"They especially love celery and lettuce, but they don't like potatoes or onions." Once again, Les strikes me as a proud father. He knows exactly what the kids will eat. "They also enjoy other fruits and vegetables, dairy, meat, chicken, and—a special treat—live fish from the hatchery at the end of the season. Lots of their food is donated by local restaurants and grocery stores." Bears are omnivores and will eat just about anything. That's a good thing, because Killisnoo and Chaik each require thirty to forty pounds of food every day.

Les shows me how, with hand and body movements, he has trained Chaik to stand up or lie down. "It used to be that the trainers worked the bears," he says. "But now the bears work the trainers—they will make the response that gets them food. Toby, the only female here, really likes to eat. She was the first to learn to signal for more food, by putting her paws together in front of her chest."

"Namaste?" I ask. It seems the bears use the Namaste gesture we Californians all learned in yoga class—palms together in front of our chest, a prayerful reminder of the divine spark within each of us—to ask for food. As if on cue, Toby begins signaling for more food.

Les says I can feed her if I want to. I have never fed a bear before. It seems like a risky undertaking, but Les seems to think I'll enjoy it, and he leads me to a hefty metal gate that opens into the animals' enclosure. It's made of exceptionally heavy-duty steel grating, painted aqua, with a space about three inches high between the ground—which is muddy from a recent rain—and the bottom of the gate. It is through this slim slot that I am to deliver apple slices to Toby.

Toby knows what's up. She wastes no time, slipping her huge paw under the door. Her claws are gigantic. The smallest one is larger than my apple slice.

Even though the fence and gate are more than enough to keep me safe, I'm scared—so scared I almost drop the apple slice into the mud. *Does mud taste bad to Toby? Will she be mad at me if I get her apple dirty?*

Toby reaches for the apple. I'm surprised at her finesse, her agility. It's almost as though she is picking up a grain of rice with chopsticks—her touch is that delicate—and I have a new appreciation of her abilities. *Namaste*, I smile to myself. No wonder the Native Americans wore bear claw necklaces to signify not only hunting prowess but also the subtler powers of spirit. Then Toby pulls the apple slice under the gate, dragging it through the mud. She is still a wild animal. And in a flash, the food is gone.

Toby's swift paw stays with me long after I leave. I came close enough to admire her delicacy—and to remember how little distance there is between reverence and risk.

Awestruck and Inspired

The bird I had come to find—the tufted coquette—
looks like a flying jewel.

My Little Red Coquette

Infatuation hums in the rainforest

A broad, wooden verandah at one end of the rainforest lodge opens onto a sheltered lawn that's lush with flowering bushes and brightly colored birds. I'm at one of my favorite spots in the world, the Asa Wright Nature Center in Trinidad's forested highlands, where you can easily see twenty different avian species in the space of twenty minutes. An ardent birder, I love them all, but my aspiration is to catch a glimpse of *Lophornis ornatus*, the tufted coquette.

I'm not the only one charmed by this little beauty; it's one of the center's star attractions. *Ornatus* refers to the bird's ornate plumage. Like many other hummingbirds, its head, back, and wings are bright green; its tail and wingtips, darker. A pale buff bar crosses the rump. But its most distinguishing features are the male's brilliant vermillion collar and crest—and the adorable black polka dots on tufts along the sides of his neck. It is an exquisite creature, and to a bird lover, catching a glimpse of one is the equivalent of coming upon a rare treasure.

Trinidad is one of the best places in the world for watching birds. More than 400 species have been identified on the island, which is about the size of the state of Delaware. Although small, it has some 40% as many species as the entire United States. But it isn't just Trinidad's species count that's impressive; it's the island's absolute *spectacle* of birds. The rainforest erupts with colorful honeycreepers, orioles, tanagers, and some of the world's most beautiful hummingbirds, easy to identify because the names are conveniently descriptive.

The copper-rumped hummer has an emerald face and throat, a dark tail, white "socks" ... and a copper rump. It's the most common visitor here at the feeders, and its territorial behavior is fun to watch. Equally territorial is the male white-necked jacobin, with its gorgeously iridescent dark blue hood in sharp contrast to a bright white belly. And yes, he has a white collar at the back of his neck. The rufous-breasted hermit has a rufous or reddish-bronze throat and paler belly. Its bright chestnut outer tail feathers are tipped in an inky black, then tipped again with a sliver of stark white. Its long beak turns downward, and the two mandibles are different colors: dark gray on top, yellow for the bottom one. When perched, he wags his tail up and down. One of the most inquisitive birds here, the rufous-breasted hermit comes up to within three feet of my face and hovers, observing me as I observe him.

While I'm watching, a spectacled thrush creeps in, meowing like a cat. Black-and-white bananaquits flit in the trees, showing off their bright yellow bellies. Time stands still, and I realize I've been blissing out with these magical little beings for more than an hour.

The center is in Trinidad's northern mountains. Once you've flown to the capital city, Port of Spain, you need to drive for several hours—much of the way on bumpy, rutted roads. My trips to Asa Wright are always worth the trouble, though; it's a heavenly spot. The canopy, in some places, is 150 feet high.

I had already made a couple of excursions from my Asa Wright basecamp. At Caroni Swamp on the island's northwestern coast I climbed into a canoe and navigated through several miles of swampland—dark, dank, and filled with mosquitoes—to see roseate spoonbills. Two hundred or so of the crimson beauties weighed down the branches of a mangrove like bright red ornaments decorating a magical Christmas tree. This alone is a world-class birding experience, but there was much more to come.

Later, driving along a road called John Boodoo

Trace south of the Arima Forest, I spotted what looked like a mass of brown, hairy tube socks hanging from a tall tree. As much as four feet long and six or seven inches wide at the bulbous base, these baskets are handiwork of the crested oropendola, a colonial breeder that weaves nests by the dozens and suspends them from the very tips of branches. I stared, hypnotized, at the swaying nests and the oropendola themselves—big black birds with a bright yellow tail and beak and striking blue eyes.

Back at Asa Wright, I head out on the promisingly named Bellbird Trail, one of several footpaths leading out from the lodge. An easy hike takes me down into deep forest where torch ginger and lobster-claw heliconia, with their bright red and orange flowers, line the pathway. Nutmeg trees, originally from South America but long ago naturalized here, grow behind them, twenty feet tall with tiny yellow flowers. Strangler figs and monkey-ladder vines twist their way skyward. Less light filters through the treetops as the overstory thickens, and the dimness enhances my other senses.

Already the day is hot and humid; moisture in the air mingles with my perspiration. My feet thud an odd, hollow rhythm on the hard dirt path. Each time a breeze wafts through, long stalks of bamboo sway, complaining with languorous sighs. Buzzing insects and melodic birdsong merge in an auditory vapor; I suspect the musicality of the Trinidadian language was born from these lilting rainforest tones.

Overhead, trees echo with the squawk of channel-billed toucans. Lots of them wing their way through the jungle, but the vegetation is so dense you can't see the toucans—you can only hear their harsh, single-syllable call. My field guide describes it as a "high-pitched *pyok*, repeated at intervals of a few seconds; at a distance [it] resembles the yelping bark of a puppy." I can't help but wonder how magical the jungle must have seemed to the author, filled as it was with the sound of flying puppies.

If you *could* see these birds, you'd recognize the toucan often kept as a pet: about twenty inches long, with a mostly-black body, a way-big beak, and a gorgeous, bright orangey-yellow breast. As if their looks weren't enchanting enough, my guidebook goes on to describe their courtship behavior: "The male feeds the female, handing berries in its ungainly bill with remarkable dexterity." So sweet.

I still use my 1980 edition of ffrench's (sic) *A Guide to the Birds of Trinidad and Tobago*—the only comprehensive volume on local birds when I bought it in 1987. It's a hardback, and nearly two inches thick. Nevertheless, it's been worth lugging around, as it's bursting with all kinds of fascinating information, not to mention the fact that the author's unusual name, Richard P. ffrench—with its double-f and lack of a capital letter—is so intriguing I often find myself wondering about the man. I don't have a regular birding buddy in Trinidad, so Mr. ffrench has become my companion.

I wander deeper into the forest, hoping to sight the bearded bellbird I've heard from the lodge. It's a light-colored, foot-long bird with a dark brown head and wings, and should be easy to spot. The male has an unusual brown "beard" consisting of dark, string-like wattles that hang down from its bare throat. Suddenly I hear the bellbird's echoing call—which sounds somewhat like a bell, hence the name. It beckons, but confuses me. One minute it seems to be nearly overhead; the next it's far away. Is it one bird, or two? Is the bird here, or there?

Turns out it's both. The male cries out with a metallic *bawk* to defend his calling territory. It has an odd resonance that reverberates through the forest like a hammer hitting an anvil. I consult my ffrench's and learn that the male regularly turns "to face a new direction, and he tends to call when a neighbor is nearby." No wonder I'm baffled. I'm apparently in the midst of dueling male ventriloquists, each turning to project his voice in a circle around his territory. The loudest caller gets the female, assuming he already has his beard. No beard, no chance. I finally find the male, perched on a branch perhaps thirty-five feet above the trail, and am surprised that he lets me get close enough to clearly see his magnificent wattles. I'm disappointed not to witness the bellbird's display preening, in which he "shows off ... a patch of bare skin on the thigh." Or so Mr. ffrench says.

A branching trail leads to the reserve's famous oilbirds, which I hear long before I see them. Screams and squawks fill the air, along with the distinct sound of vomiting. It's an awful racket, accompanied by a nasty odor. I scramble over rocks and stop a little ways outside the entrance to the dim cave.

Up at the lodge, someone said these creatures are called oilbirds because so much of their body is fat, and that they were once commonly speared on the end of a sturdy stick by indigenous people and lighted on fire to form a torch. Consequently, there aren't many left—but the colony here is flourishing. In fact, one of the original reasons the Asa Wright Nature Center was founded was to help protect these rare birds.

I consult Mr. ffrench for an explanation. The oilbirds, he says, were used by locals as a source of fuel for cooking and light. No specific mention of the torches. He also says they are nocturnal, eating fruit they pluck from trees in flight, at night, which they're able to do because they navigate like bats using echolocation. They later disgorge the seeds and use the resulting vomit to make their nests, which are "constructed mainly of regurgitated fruit forming a paste." Ah—that explains the noise, and the smell.

This cave is not a pleasant place to visit. In addition to the incessant retching and strong odor that accompanies it, there's the matter of my having disturbed the birds, even though I'm keeping my distance. They're huge—with three-and-a-half-foot wingspans—which is particularly distressing when they are flapping restlessly in the confines of their narrow, cramped cave. Skipping the cavernous chaos, I spare the birds and myself any further torment and head back for the trail.

I'm hoping for an antidote to the chaotic oilbird experience, and find it in the golden-headed manakin lek that's right along the path. A lek is an area where male birds gather to compete with one another in their mating rituals. These dancing birds are only about three-and-a-half inches long, with tails so short they're barely noticeable. They sport jet-black bodies, bright yellow heads, a short pink beak, and pink legs. A white ring circles each eye, giving the impression of bespectacled intelligence—until you see the males' strange, Michael Jackson-moonwalk-like mating dance. It's such a silly show that it gives me the giggles, and I need to move along so I don't scare the birds away.

But the afternoon is sweltering, so I'm happy to head back to the lodge where I can relax with a rum punch and watch the hummingbirds buzz around the feeders. I've been here for two days, but still haven't seen my little red coquette. After searching for so long, I think of the bird as "mine," in the same way my mother-in-law, who had a crush on the actor James Garner, always refered to him as "my" James Garner. Unilateral infatuation has made the little red coquette *mine*.

Out on the verandah, the air is filled with sweet, high-pitched tweets; feathery flashes of red, topaz, sapphire, green, and purple zip by. They call to mind Einstein's Special Theory of Relativity, which suggests that time slows down as speed increases. What does that mean for these tiny birds, whose hearts beat some 1,200 times each minute? Even their wings vibrate faster than the human eye can see—up to eighty times a second.

I try to imagine what I must look like to the rufous-breasted hermit as it hovers so close I feel the rush of air from its wings. Does time bend when it flies? Do I appear to move slowly when the bird is whizzing through the air, but more quickly when it's hovering? Do I seem barely animate, like trees look to humans?

The sun begins to sink, and the hummers are off to bed. I'm up early the next morning, sitting at the far end of the porch, just across from a big bush full of red flowers. I've heard it's the tufted coquette's favorite spot. Rather than venturing into the bushes, I sit as quietly as possible; this is the best way to sight a shy hummer.

The other birds fly in and out: yellow-and-black orioles; purple and green honeycreepers; cocoa woodcreepers; gray-fronted doves; an assortment of blue-gray, bay-headed, white-shouldered, and silver-beaked tanagers; the purple-backed and famously musical violaceous euphonia with its sunny yellow chest and forehead. Hummingbirds are also out in full force: white-chested emeralds, black-throated mangoes, blue-chinned sapphires, long-billed starthroats.

Finally, around 9:30 in the morning, the female tufted coquette appears. She's so tiny and quick that at first I mistake her for a giant bumblebee. At just over two-and-a-half inches—including her tail—she's one of the world's smallest birds. She's paler than a male and lacks some of his distinctive field marks, but her tiny size and a prominent white band across her rump make identification easy. She buzzes from flower to flower, never staying long in one place, and spends altogether too much time hidden on the far side of the bush—a true coquette.

Then the male appears and honest-to-gosh, my heart catches. He is spectacular: iridescent green, with a magnificent vermillion crest that pokes up nearly an inch from his forehead like a bright crown. Long, pale scarlet plumes stretch backwards from his cheeks and over his shoulders. The feathers are of varying lengths and each is tipped with a dark green-black splotch, giving the tufts their polka-dotted look. He can flatten the tufts in against his body or project them out from the sides of his neck in an exquisite display—which he does several times, hovering just a few feet away. For several seconds I am overwhelmed by his beauty.

Finally, I relax. I have hunted down the roseate spoonbill colony and the elusive bellbird. I watched the golden-headed manakins dance and heard the rare oilbirds retch. I enjoyed the magic of sitting still as sparkling hummingbirds hovered silently, or whizzed by, or chased competitors away from the feeder. And I learned an exhilarating lesson about birdwatching and life: Sometimes, if you put yourself in the right place at the right time, and have a bit of patience, you don't have to search for treasure—it will fly right up to you and say hello.

The wildebeest, also known as the gnu, is surely one of God's oddest creatures. But they're a keystone species—absolutely essential for the health of their entire ecosystem.

Who Gnu?

A surprising case of animal intelligence
on the Maasai Mara

G stands for Gnu, whose weapons of defence
Are long, sharp, curling horns, and common-sense.
— Hilaire Belloc, *A Moral Alphabet*

I have been fascinated by the gnu from an early age, doubtless because of its prominent place—the seventh letter—in my favorite childhood reader, *The Bad Child's Book of Beasts and More Beasts for Worse Children and a Moral Alphabet*. Originally published as three separate books in the nineteenth century, the trio were combined and reissued in 1961. I have no idea why my parents, who had a modern approach to life, thought it was appropriate for me to learn to read from a book written in the previous century with "Bad Child" and "Worse Children" in the title, but they did, and I did, and there you have it: I still own the well-worn reader, and I still feel a special affinity for that particular beast.

I had seen the gnu as a child, of course. It was on *Mutual of Omaha's Wild Kingdom*, which I waited for eagerly every Sunday morning. Watching faithfully week after week as Marlin Perkins presented his cavalcade of animals—each stranger and more wonderful than the last—I hoped that one day I'd see them in real life. The gnu, Mr. Perkins explained, is a member of the antelope family, and earned its Afrikaans nickname—wild beast, or *wildebeest*—for the frightful appearance of its oversized head and neck, shaggy beard, and sharp horns.

But what the wildebeest lack in attractiveness they make up for in fortitude: Every year they lead the Great Migration in a circle that stretches over a thousand miles of African savanna. Zebra, gazelle, giraffe, and many other animals join in, sometimes plodding, sometimes running, and occasionally even swimming through crocodile-infested waters, all driven to find the new grasses that sprout with seasonal rains. Stealthy predators follow in pursuit, making this the earth's greatest movement of land mammals.

It's an epic trek—Mother Nature at her biggest, an extravaganza that puts humans in their proper place. And the wildebeest is the star of the show. I have always wanted to witness it.

I discovered that the best way to view the migration is actually from the air—specifically, from a hot-air balloon. It positions one close enough for a good view of the individual animals, yet high enough to be out of the way—this is very important, considering they're prone to stampede!—and distant enough to appreciate the majesty of two million animals, plain and patterned, predator and prey, all on parade.

Which is why, when I visited Kenya, a balloon ride over the wildebeest migration was at the top of my to-do list, despite the fact that it required me to put my trust in a sphere of the thinnest nylon—with a big hole in the bottom—and burners running on the same propane fuel as my home barbecue grill. My guide for the experience would be Captain Tom, an expert on African wildlife as well as on piloting an air-driven craft with only two burners and two holes—an upper vent and a lower one—as controls.

We began in the cool of pre-dawn, when light was just beginning to tint the sky. The ground crew laid out our balloon—a bright yellow 80-foot bubble, striped with red and blue—turned on a fan, and opened the blast valve, releasing a rush of blinding flame that roared from the burner and gradually filled the balloon with the heated air that would carry us skyward. Dangling beneath was a wicker gondola about the size of a five-person hot tub, with seating for ten. Despite recent advances in technology, old-fashioned wicker is still the material of choice for the basket because it's lightweight yet durable enough to withstand the inevitable bumpy landings without much damage. I climbed aboard with Captain Tom and five other guests.

At first nothing much seemed to be happening; the balloon remained low and we couldn't see any animals. I was beginning to wonder whether we were in the right place.

"Are we going to go up a little higher?" I asked, hopeful.

"Those small hills are blocking the breeze," Tom explained. "Things'll get easier as soon as we gain a little altitude."

As we rose, the dim morning light revealed perpetual plains and low, rolling hills covered in golden grasses. Occasional small trees dotted the savanna, giving it a playful, patterned appearance. This was the *Wild Kingdom* I had dreamed about as a child, the Africa of Beryl Markham and Isak Dinesen I had read about as a teenager, the home of the Great Migration. This was the Africa I had longed to see!

As we peered into the dawn, eager for our first look, Tom filled us in on the lives of the animals below. "The wildebeest are active both day and night; they're constantly grazing," he explained. "They calve in late January or February. Well over a quarter of a million are born in about three weeks."

I did the math. "That's five hundred babies per hour—can that be right?"

"Yes," Tom agreed. "It comes out to about that. The once-a-year glut for predators makes it so more newborn wildebeest can survive their crucial first few weeks."

That reproductive rate says a lot about the perils of being a wildebeest: Although the population of about one-and-a-half million is relatively stable, they need to replace twenty percent of the herd every year. These beasts are nothing if not tough, though. The calves learn to stand faster than any other mammal, can walk within ten minutes of their birth and can run fast enough to keep up with the adult herd when they're only a few days old.

In early July, when I visited, the migration was in full swing. "There we go," Tom said as he navigated the balloon toward still-invisible migrants. "Look over there—a dozen Thompson's gazelle." The gazelle were spooked by the roar of the balloon's burner and scattered chaotically, bounding and zigzagging beneath us like frightened rabbits. From above, it was impossible not to notice the imbalance: our curiosity announced by fire and noise, their response immediate and involuntary.

"Wildebeest facilitate the migration of other grazers,"

Tom explained. "The general route is shaped by the availability of food and water, and the need to avoid predators. Along the way, the wildebeest also find the grasses that are highest in phosphorous, which lactating females need. They also select grasses with the highest nitrogen content."

"How do they know which grasses are highest in minerals—can they taste the difference?" I asked, wondering what phosphorous and nitrogen might taste like.

"No one knows," Tom said. "After the wildebeest have grazed, 150,000 Thomson's gazelle follow along and eat the leafy, new-growth grasses stimulated by the wildebeests' foraging. It's a complicated relationship scientists are just beginning to understand."

The gigantic herd has been estimated to produce 420 tons of dung each day, and supports the livelihood of more than one hundred species of dung beetle.

The sun brightened and soon we could see wildebeest below, tens of thousands of them, stretching out to forever, impossible to count. In some places they gathered in amorphous masses, grazing lazily, but below us they formed an endless single-file line, marching off the horizon to eternity.

We floated downward until we could hear the wildebeest calling out, high-pitched and whiny, a giant herd of snorting, bawling cattle. Sometimes they croaked like frogs. Occasionally the animals moaned in what sounded like pain or ecstasy—or perhaps relief, given their grassy, gassy diet. Drifting lower still, we got a good look at the beasts, which have comically spindly legs and manes that often sprout abruptly from their necks like unkempt Mohawk hairdos.

The tranquility was suddenly replaced with a monstrous roar and a belch of fire as our hot-air balloon descended a little too low. Over and over again, Captain Tom floated us down to just above the treetops, then fired up the jets and whooshed us higher for another heart-stopping view. "I hope we'll see a lion today," he murmured, and within seconds a lone female appeared, flying through the tall grasses, too low for the wildebeest to see. She was heading straight for the long line of wildebeest, perpendicular to their trajectory as they marched across her field of vision, blissfully unaware, like toy ducks moving across the back wall of a carnival shooting gallery.

Wildebeest are sometimes called "the clowns of the savanna" because of their tendency to run in circles or otherwise cavort for no apparent reason.

I wasn't sure I wanted to see this part of the cycle of life up close, and was relieved when Tom observed that she wasn't hunting. "She's by herself. If she was after food, she'd be with a small group." The wildebeest must have seen her then, or heard her. They stopped walking and peered in the lion's direction. Several even ventured a few steps toward the lion with what looked like myopic curiosity.

Not a brilliant strategy, in my opinion—why didn't they keep moving? Finally deciding danger was indeed afoot, the wildebeest took off in a wild gallop. But they stayed in their single-file line, still moving perpendicular to the approaching lion. Why didn't they turn and run away? Was this the reason they had to have so many calves each year? "Do the wildebeest have any way of protecting themselves from predators?" I asked, as the lioness veered off. She was apparently just out for some exercise, and not in the mood to take on a tackle.

"They can run at speeds approaching eighty kilometers an hour," Tom pointed out, "and the males can weigh five or six hundred pounds. They can inflict considerable harm, even to a lion. But their primary protection comes from gathering together in large herds. That way there are plenty of eyes to watch for predators, and the young animals are protected by the older, larger ones."

I later read that wildebeest use other cooperative behaviors, such as taking turns sleeping while others stand guard against a night attack. They even seem to understand the calls of other species, responding more strongly to baboons' alarm barks than to their other vocalizations.

Sadly, we saw just a short snippet of the Great Migration. Our balloon ride, unlike the journey of the wildebeest, lasted only an hour. When we came down it was with a bang. The descent was gentle but fast, and the basket bumped along on the ground, then tipped onto its side and rolled, dragging through the grass for eighty feet or so before it deposited us in a jumble of arms and legs.

No one was hurt, and we crawled out of the basket and walked a short distance to where the ground crew had prepared an elegant champagne breakfast. Chafing dishes held hot bacon and scrambled eggs, the scents of warm cinnamon rolls and Kenyan coffee wafted through the air, and sunrise-colored fruit juices shone in the golden morning light.

My luxury balloon adventure couldn't have been a sharper contrast to the amazing survival spectacle I'd just observed. Wildebeest travel a thousand miles a year, every year, without a map or GPS. They know how to find the most nutritious grasses. They work together and, in spite of fierce predators, manage to survive in impressive numbers. I raised my champagne glass in a heartfelt toast to these creatures, who are certainly smarter than they first appear.

Who gnu?

A leatherback turtle thinking about laying her eggs.
This is what the beach at Matura would have looked like,
if there had been any light at all when I visited.

Leatherback Love

Long nights with nesting sea turtles
on a Trinidadian beach

Matura Beach is dark tonight. I can't make out much, except the faraway glow of a high-rise hotel a mile or so away, but I'm hoping to see a leatherback turtle. The females come here by the hundreds every year to lay their eggs, depositing up to 120 at a time in nests the mothers dig into the soft brown sand.

I'm glad to be with my guide, Francis, because I'm pretty sure I'll need him to lead me back to the footpath that winds through mounds of low brush and sand back to my van in the unlit parking lot behind the beach.

Matura Beach is on the eastern shore of Trinidad, the southernmost island of the West Indies, which lies just seven miles offshore from Venezuela. It's one of the best places in the world to see the leatherback turtle, *Dermochelys coriacea*. At this time of year—

early May—there could be fifteen to twenty turtles laying eggs on the beach each night. By the peak of the season in June, there might be as many as 300 female turtles, each digging a nest and depositing her eggs. Sometimes they accidentally dig up another nest in the process.

You'd think the turtles would know enough to spread out, but that's no longer an option because there isn't much good habitat left. These giant sea creatures—which can reach up to 1500 pounds—require dark, secluded beaches adjacent to deep water for successful nesting. They rely on light to aid their navigation, crawling in from the ocean, which sparkles dimly in the starlight, to areas of dark vegetation on the beach. There they lay their eggs in shadow nests, lumbering back to the lighter-looking ocean after covering the eggs with sand. The baby turtles, vulnerable to predation and poaching by hawks and other animals—as well as humans—are safe in the darkness. This is their incubator; their nursery is the deep water nearby where offshore reefs cannot injure their still-tender bodies.

At eight o'clock in the evening it's still a little early for the turtles, but Francis is not resting. Lean and full of energy, he shouts over the ocean's roar about the life of a leatherback. "People say the turtles cry, but they don't," he hollers. "They eat a lot of jellyfish—

which are very salty—and their tear glands excrete the excess salt."

Francis, who has lived on the island of Trinidad all his life, speaks in a lively and melodic pidgin. "When they come outta the sand, it's like a volcano erupting. The first ones out wait for the others," he explains. This is only one of the clever survival tricks the leatherbacks employ: When they're all heading for the water at the same time, fewer are picked off by predators. Francis tells me about some of their other adaptive behaviors. "One female mates with five to seven males and stores up all the sperm. Sometimes a first-time mother will have some problems, but if she can't lay the eggs on her first try, she'll save them up and come back another night."

"What kinds of problems can she have?"

"She might not find the right spot to lay her eggs. The temperature of the sand determines the sex of the baby turtles: You got cool guys or hot chicks. Easy to remember." He's referring to the fact that the temperature of the sand they're laid in determines whether the eggs will hatch into male or female babies. "Look! There's Orion, and the Southern Cross. The stars are bright tonight."

Francis is right; the stars are getting brighter by the minute. Soon there is enough light that I can just make out a turtle-sized shadow on the shore. At least I think it's a turtle. It's large and dark and barely moving. "There she is; she's making her way up!" he exclaims. "She is coming up from the surf. Few people see this!"

Fifteen minutes later the turtle has made her way about twenty feet up the gentle, sandy slope, but then she turns and heads back toward the ocean. Francis places himself strategically between the turtle and the shoreline, encouraging her to try again, and once more she heads across the sand. "Yes, she is a first-timer," he announces. "She doesn't know how to do it yet."

Climbing the low dunes and finding the right spot is an agonizingly slow process, and not very exciting to watch. "She is looking for the temperature of the sand," Francis explains. I'm shivering in the cold wind, and am beginning to think about calling it quits when the leatherback finally turns to face the ocean. "She's moving the flipper; that's a good sign. You are gonna see eggs tonight. Once she's settled we'll see everything!"

Our turtle begins to dig, alternating her left and right flippers. "She is gonna start excavating. See, when she is going left and right, she has started." This, too, is a very slow process.

"How deep will she dig?" I ask, wondering how long we're going to be on the beach.

"See, she digs with her back flippers. So she's gonna dig deep. She'll dig out a lightbulb-shaped hole about two or three feet deep to deposit her eggs in. It's as deep as her flippers can go." Then Francis sees a problem. "She's digging on top of an existing nest, and the eggs that are already there are getting pushed aside."

"What will happen to the eggs from the first nest?" I wonder aloud.

"This is not good," Francis explains. "The eggs are gonna die. Bacteria will spread into the nest environment. The clutch will not hatch, but this is nature takin' its course." As though she understood him, our turtle stops digging and moves on. Finally she finds a suitable spot on the sand, digs her nest, and begins laying the eggs.

"There will be two types of eggs, fertile and non-fertile; the non-fertile ones make roomspace in the nest." Francis turns on his red-lensed flashlight and shines it on the hole she has excavated. We can see the eggs, which look like golf balls, plopping out of the turtle and falling down into the hole. Sometimes they come out one or two at a time; sometimes seven or eight come together. Some eggs are a little smaller than the others; these take up space—or "roomspace," as Francis puts it. I imagine they help regulate the temperature of the nest. "The shell is rubbery-like," Francis explains. "It never gets hard like a chicken egg."

After twenty minutes the turtle is finished, and appears to rest. "Now you can turn on your flashlight—she is still in her hypnotic state. You can touch her if you want to," Francis offers, explaining that while the leatherback is in this short-lived "hypnotic state" we can touch her without fear of bothering her or interfering with the reproductive process.

I reach down and stroke the turtle's flipper, which is surprisingly soft. But that's not my only surprise. Something about having watched this undertaking makes me feel close to the turtle, and invested in the outcome of her reproductive rite.

"How many of these eggs will hatch?" I ask, suddenly feeling protective.

"There's only a one-in-a-thousand survival rate, and they cannot be reared in captivity," Francis explains. "That's why it's so important for us to help the turtles."

A second guide joins Francis, and they measure the turtle's shell—this one is 117 centimeters, or just under four feet, from side to side. They also check for metal tags on her hind flippers, and, finding none, clamp one on each side. She was, indeed, a first-timer. The tags will identify her next year.

Francis detects some invisible sign and announces that our time is up. "She is finished now. Turn off your flashlight." We're back to watching by the meager glow of Francis's red beam.

He was right; the turtle has finished her short rest and begins to cover up the eggs by flipping sand into the hole. When it's adequately covered, she crawls back to the water.

Soon enough we'll be crawling back to our beds. I can't imagine being out here in the wind and cold night after night—even the turtles lay their eggs and then take off. I ask Francis how long he's been coming to the beach and why he keeps at it.

"I do this for future generations, and to preserve the turtles so we don't lose them," Francis explains. He clearly adores the leatherbacks. "I've been a guide at Matura Beach for more than twenty years. I even come here on my night off; I love to be here."

I nod in the darkness. And I get it: the stars shimmering over a black sea, the rush of wind and tides, the slow crawl of generations. As we head back to the parking lot, a strange feeling comes over me; I think it's leatherback love.

It's a privilege to see this fierce beauty up close.

The Great Bald Eagle Fly-In at Klamath Basin

Snow, scavengers, and sex in the sky

Each winter, just as surely as frost laces the trees and ice begins to silence the wetlands, hundreds of bald eagles congregate in a "fly-in" at northern California's Klamath Basin. The best time to see them is during the mid-February peak, when as many as five hundred of the raptors gather there. The weather is numbing.

The fly-in is supposed to be one of the most spectacular events a wildlife enthusiast will ever witness. An enthusiast, that is, who doesn't mind driving hundreds of miles from her cozy home in the San Francisco Bay Area into the boondocks at the Oregon border, or staying overnight in an unpainted cinderblock duck-hunting lodge with a teeny, tiny, mostly ineffectual heater and no ensuite bathroom. One who is willing to rise before dawn and brave subfreezing temperatures.

That wildlife enthusiast would be me.

Bald eagles are awe-inspiring creatures. Their eyesight is said to be so sharp they can see a fish from a mile away, or a rabbit from two miles. They easily climb to 10,000 feet in the air, and reach speeds of one hundred miles an hour in a dive. They can carry the heaviest load of any bird.

We haven't always treated the bald eagles well, despite their raw power and fierce behavior—and not to mention that they're our national emblem. In the early eighteenth century they were a common sight on much of the continent, where as many as half a million lived.

Then we shot them. Trapped them. Poisoned them. We dismantled their habitat and tainted their food, and finally laced the environment with DDT, a chemical that quietly robbed them of the ability to reproduce. By the middle of the twentieth century, the symbol of American strength was nearly gone. Fewer than four hundred nesting pairs remained in the lower forty-eight states.

Recovery came slowly, and not by accident. Laws were passed. Habitats protected. DDT was banned. Only then did bald eagles begin to reclaim the skies we had emptied. Today, roughly ten thousand nesting pairs live in the lower forty-eight states—a number that still astonishes, given how close we came to losing them. In winter, the Klamath Basin becomes one of the best places to witness that improbable renaissance, when the air itself seems to fill again with wings.

So one icy day in early February I drove north to the Lower Klamath National Wildlife Refuge near the California-Oregon border. Established by President Theodore Roosevelt in 1908 as the nation's first waterfowl refuge, the reserve is now a National Historic Landmark. More than three hundred avian species—ducks, geese, swans, cranes, other waterfowl, and shorebirds—wing their way through, part of the fall migration of one million birds. As I drove, the sky occasionally darkened with flocks of honking geese.

The Klamath Basin, an immense expanse of wetlands, meadows, shallow lakes, and craggy cliffs, seems breathtakingly vast and fantastically ferocious at this time of year. Remains of carcasses—it's rarely clear what animal they once were—litter the snow-covered grassland. Eagles huddle in frozen fields, others soar skyward, still others dive from the heavens. The birds are enthralling. Often immature eagles, being less able hunters, try to steal scraps from older, more experienced birds. It's easy to ID the immature birds by plumage as well as behavior; bald eagles don't develop their distinctive white head and tail feathers until they are four or five years old. Eventually these young birds will need to learn to catch their own food if they're going to make it through the year.

I was eager to witness the eagles' unforgettable courtship ritual. A pair will climb so high they are nearly invisible, circling one another with elaborate swoops and chases. Then the two birds lock talons and cartwheel head-over-tail, plummeting in a death-defying freefall. At the very last moment before crashing to earth, they separate and glide off in apparent nonchalance. When an observer is able to breathe again, it's generally with an awestruck gasp. The display is amazing.

In fact, everything about eagles is amazing. I'm not the only person to think so—the fierce raptor was emblematic for the Babylonian, Egyptian, and Roman empires. Christian, Aztec, and Muslim texts include it as a spiritual symbol. Eagles represent courage, strength, and vision for many Native Americans. And the bald eagle has been the national emblem of the United States since 1782.

Eagles still fascinate people all around the world. A dozen raptor fanatics stayed in the spartan duck-hunting lodge that night. Unlike me, the others were not first-timers. Most had driven up from the San Francisco Bay Area, but one couple was from Europe. There's something irresistible about eagles' power and grace that attracts birdwatchers year after year.

The next morning we gulped hot coffee together in

the parking lot before heading out. Tufts of dead grass crunched beneath our feet, ice coated the pavement, and snow blanketed the hillsides. Our breath turned to clouds in the frigid air. It was a daunting start, but I was game. Locals say the colder the weather, the more eagles there are.

Up well before sunrise, we drove miles of back roads through early-morning darkness to reach a particular spot we had been told was the eagles' night roost, a deciduous forest on a sheltered slope. The bare-branched trees provide easy perching, roosting, takeoffs, and landings for eagles that cluster here at night.

Our source had been a good one. Stark black skeleton-like trees stood outlined against a pale hillside. And on their branches, shadows—the promise of what we had come to see. We didn't need daylight to ID them. The dark brown bodies and snow-white head-and-tail feathers of adults, along with their impressive size—average wingspans approach seven feet—made the raptors unmistakable.

Slowly, the dim light brightened. One bird rose, then another. Soon hundreds of majestic eagles ascended in the thin post-dawn air. I watched in awe as they labored silently upward one by one, then wheeled together in a "kettle" of more than fifty individuals, spiraling in a lazy vortex on a rising thermal. They were surveying the land below, looking for food.

Dozens more loitered in treetops and on the ground less than one hundred yards from the road, indifferent to our presence, occasionally scrapping with each other for morsels of scavenged food. Eating will be their main job for the coming weeks. Flying in from as far away as Alaska and the Yukon—some have traveled well over a thousand miles—requires a lot of energy. But the rewards of wintering in Klamath's extensive shallow marshes and grassy uplands, where nearly two million waterfowl find food and shelter, are great: fish and ducks are easy prey.

There's also the matter of the calves. Bald eagles are predators, but we've heard the raptors also work as a clean-up crew here in cattle country, feasting on nutritious afterbirth when they come across it during the bitter cold of calving season.

My hands and feet grew numb as we watched for hours. There was plenty of time to talk with other eagle-watchers, many of whom had brought along giant spotting scopes—telescope-like lenses mounted on tripods—perfect for training on one position so we could view the relatively stationary birds. I connected with one guy, Walt, with a particularly big scope and no apparent friends. Lots of birders are loners.

"That's quite a scope you've got there."

"Sure, take a look!" Walt knew what I was after.

"Thanks." I peered through the eyepiece. Walt had the scope trained on a beautiful adult with a bright white head. The eagle was ripping at what looked like the remains of a rabbit; there was surprisingly little blood. "Whoa! We're watching the circle of life here."

"Yep." Walt pulled a wad of waxed paper from the pocket of his parka, unwrapped a sandwich, and took a bite.

A man of few words, I thought, turning back to the scope. "How did you get into birding, Walt?"

"You don't want to know."

"Sure I do. Tell me the story."

What I expected was a story like my own. I've loved birds as long as I can remember; my theory is that it's because my mother hung a mobile with birds on it over my crib, and thereby imprinted me with that connection. I read that Frank Lloyd Wright's mother decorated his nursery with pictures of cathedrals, and that's why he became an architect. But Walt's story was nothing like that.

"I shot a bird once," he said.

I didn't look up from the scope. This seemed like a conversation best had at a distance—an emotional distance anyway, not eye-to-eye. But Walt did not continue.

"And?" I kept my head down. The eagle continued

ripping at the carcass, using both feet to hold it down. He ripped off fur, muscle, stringy entrails, and ate it all. He was fast.

"My next-door neighbor, Dennis, we called him Denny, had a BB gun, and we were out playing with it one day. He gave me the gun and told me to shoot a pigeon. I had never even held a gun before, and I didn't want to shoot it, but Denny said I was a sissy—no offense—if I couldn't shoot a pigeon."

"No offense taken. So you shot it?" I glanced at Walt. Suddenly, *not* looking at him seemed too intimate—as though I was making too big a deal of his story. Plenty of kids shot birds with BB guns.

"I didn't kill it, but I shot it down off the telephone wire. We ran over and looked at it, and it was lying on its side, staring up at us with one eye." Walt's voice wavered a little, and I wondered whether he had ever told this story before. I wanted to check for a wedding ring—surely he would have told his wife—but his hands were jammed into his pockets. "Denny grabbed the gun from me and shot it in the head."

"Was that traumatic?" It seemed like an obvious question, and I didn't want to avoid it.

"I never touched a gun again."

"And now you're watching eagles."

"A pigeon is a dove," Walt said, unwilling to move on. "The bird of peace." He lapsed into a world deep with memories.

After that we communed with the eagles in snowy silence: the graceful show as they rose effortlessly into the sky; the slow, dizzying spiral of their kettles; the heart-stopping eagle sex. Walt's spotting scope remained trained on the field in front of us, magnifying the astonishing power of huge, bloody beaks and strong talons as these carnivores ripped at the frozen remains of lesser creatures. I was fascinated, in a secret place in my heart, with such majestic carnage. I could only wonder what Walt was feeling.

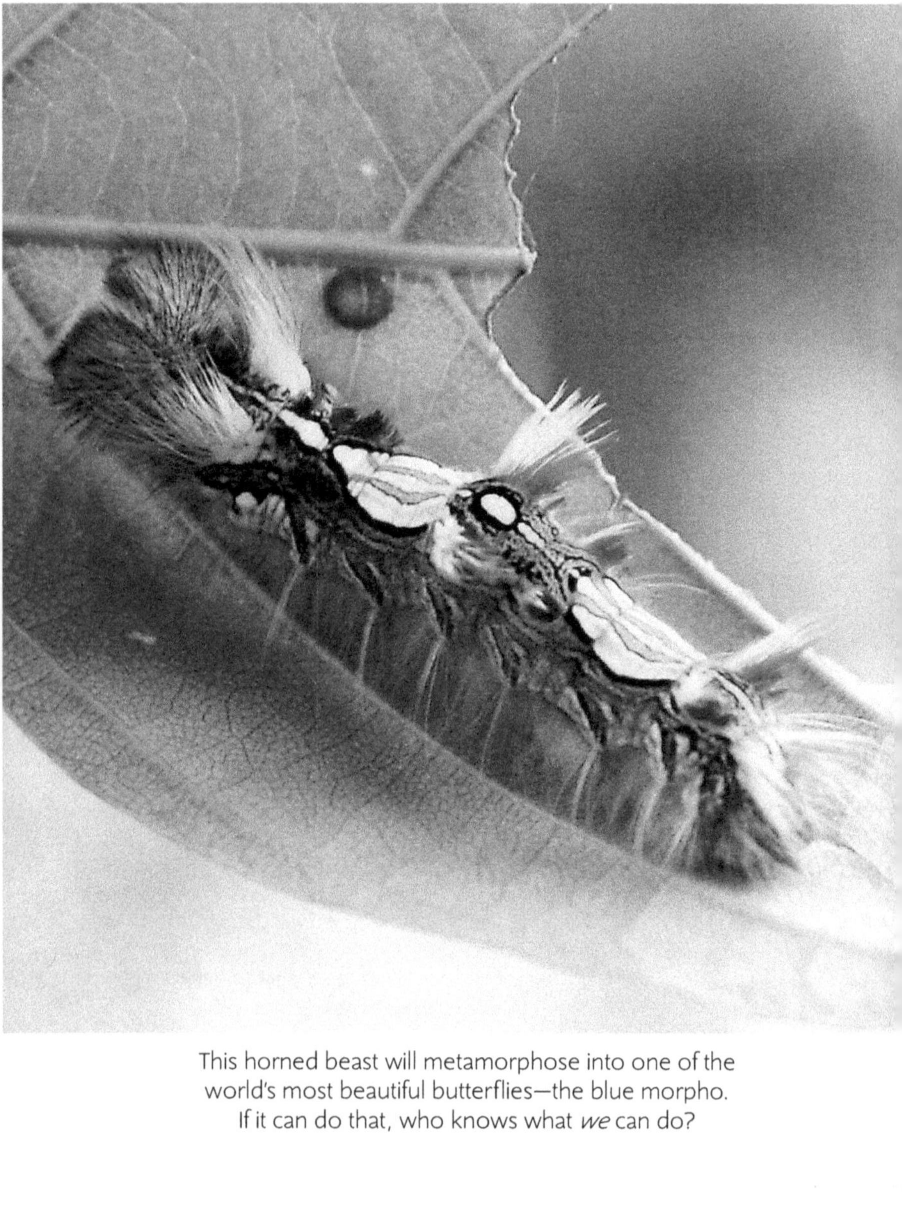

This horned beast will metamorphose into one of the world's most beautiful butterflies—the blue morpho. If it can do that, who knows what *we* can do?

Metamorphosis in La Guácima

Butterflies become a talisman for the future

Brightly colored wings explode around me like silent fireworks. Paper-thin sheets of iridescent blue, broader than my hand, float inches from my face, while orange-and-black Isabellas punctuate the shadows. Black-and-yellow swallowtails drift through the warm breeze. The effect is immersive and faintly unreal, as though I've crossed some kind of invisible threshold.

Motionless, I watch the butterflies flit from plant to plant. They aren't searching for nectar; they seem to be playing, resting, chasing each other, and mating. I'm seeing a world I don't fully understand—but my guide, Vanessa, is about to give me an education.

I'm at a butterfly farm—or *mariposario*—on the outskirts of La Guácima, a dusty village in central Costa Rica. One of the world's leading suppliers of butterfly chrysalises for live exhibits, it's aptly named The Butterfly Farm. The main greenhouse, 7,500 square feet of open air enclosed with wire mesh, keeps the butterflies in and most predators out. Narrow pathways wind through shaded jungle, dotted with sunny patches of flowering plants: passionflower, lantana, cosmos. One path leads to a sheltered grove where a small waterfall splashes into a quiet pool—this is the favorite haunt of the blue morphos.

The town of La Guácima is enthusiastic about the farm. Its streets are lined with butterfly murals painted on the sides of buildings: butterflies with flowers, butterflies with chrysalises, butterflies with graceful women.

Most of the murals are lively and charming, but one disturbs me. Across a gray wall, gigantic trompe-l'oeil beauties hang side-by-side from a long clothesline, delicately attached by their wingtips with brightly painted clothespins. Their colors drip off like water: teardrops of red, yellow, green, blue, and orange falling from flightless wings. It's a reminder that here, the butterflies are livestock. But perhaps they are also something more.

Vanessa will show me around the farm. In her early twenties, she's tall and lean, with high cheekbones and almond eyes. A tiny blue butterfly is painted on her left thumbnail, its black body flecked with silver glitter.

Matching blue butterfly earrings dance in Vanessa's long dark hair.

She holds a giant owl butterfly, her manicured thumb and forefinger pressing gently against the outsides of its two upper wings, which are closed. Overall, it's a dull brown-and-gray creature with a wingspan of eight inches, but the back wings bear dramatic eyespots—round black "pupils" inside golden "irises" that darken to orangey-brown at the edges.

"You see the eyes?" Vanessa asks. "They protect the butterfly from predators." They're on the underside, showing when the wings are folded. I imagine a small animal confronted by those staring "eyes." The wings flit and I startle; for a moment, the false eyes really do look alive.

"You see it!" Vanessa laughs. She stoops and picks up a twig the size of a sewing needle and uses it to gently unroll the butterfly's coiled proboscis—its feeding tube.

"Doesn't that hurt it?" I ask, alarmed.

"No, not if you're careful." Out and out she rolls the proboscis. Fully extended, it's more than an inch long. The butterfly doesn't seem to mind. It doesn't even twitch. "This is actually two strands that fit together like a zipper to form a drinking tube for nectar and the juice of fermented fruits. Sometimes they even drink the blood, sweat, or tears from other animals. The tips of their legs taste the plant they land on."

Vanessa releases the butterfly, and it flutters over to a feeding station, landing on a browning piece of overripe banana and extending its proboscis for a taste. "Owl butterflies are crepuscular; they're most active at dawn and dusk. During the day they hang out here at the juice bar, getting drunk."

The "juice bar" is a feeding station: a two-foot-square plywood tabletop, once painted green but now so chipped and faded it seems like part of the landscape. Cups of light red fruit juice and slices of rotting banana sit on the plywood. A couple dozen owl butterflies cover the whole thing, sipping and flapping, and fluttering their eyespots. The effect is weird and a little creepy. I understand why it scares predators away.

"They're getting sauced at the bar all day!" I observe. "Is that why they fly so erratically—because they're drunk?"

"Maybe," Vanessa allows. "But the unpredictable flight patterns help protect them from predators."

I have always thought that odd flight should have earned them the moniker "flutterby" instead of butterfly, but it's probably not realistic to hope for a change from the Old English name at this late date. "What do they do at dawn and dusk?"

"They're mating, or defending their territory."

"So they drink all day, and fight or philander in the evening—what thugs!"

We laugh and move down the path a few yards. Vanessa stops at one of the many banana trees, pulling a leaf down for us to view. Next to her thumbnail are three tiny spheres, each the size of a pinhead. "Here are some owl butterfly eggs that were laid recently," she says. I'm amazed she found these tiny eggs in such a huge space; she must know every inch of the mariposario.

"These eggs will darken, but they won't get any bigger before the caterpillars emerge," Vanessa says. "The caterpillar's own eggshell is its first meal."

Hmmm. That's also a little creepy.

She moves on to another banana plant and pulls down a long green leaf. Along its central spine are seven small caterpillars, each the same light-green color and smooth texture as the spine. The smallest ones are about a quarter of an inch long; the larger ones are close to an inch. "Newly hatched caterpillars are preyed on by lizards, frogs, birds, monkeys, and snakes. Even the ants eat them," Vanessa says. "These are camouflaged with brown splotches that resemble damaged leaves. Predators tend to overlook them."

Next, she points out another batch of caterpillars. Six inches long, these are the same species, but look nothing like the little wigglers we just saw. They've changed from small, green and smooth to big, brown and hairy—and they've grown spiny horns on their backs and weird projections on their heads that make them look like tiny triceratops dinosaurs. "See how they're camouflaged to resemble the bark of a tree? They'll stay in the caterpillar stage for two months, eating their host plant and growing quickly. They prefer plantains and bananas; farmers consider them a pest."

The Butterfly Farm—the first in Latin America—was founded in 1984 by American Peace Corps volunteer Joris Brinckerhoff and his wife, Maria Sabido. Brinckerhoff had degrees in economics and political science. At the time, he knew nothing about butterflies or entomology, but he'd had a chance encounter with a butterfly enthusiast and was also inspired by E. F. Schumacher's book, *Small is Beautiful*. Butterfly farming, Brinckerhoff thought, would be a good way for local families to generate income without causing ecological damage.

Over the years, Joris and Maria learned a lot: to

identify butterflies; to understand their various life-cycles, habitat, and food requirements; to create breeding methodologies; and how to collect and ship pupae so the butterflies had the best chance of survival. They learned how the weather affects butterfly reproduction: Too little rain means the butterflies won't lay eggs; too much rain and the eggs won't survive. They also learned about the complexities of government and airline regulations—shipping chrysalises is complicated.

These days, the business is thriving. Instructors use it as a case study for Harvard MBA candidates. The farm is one of the leading suppliers of butterflies for live exhibits, exporting 400,000 chrysalises a year to 120 clients worldwide, including the Smithsonian. But they aren't all raised here. At a price of $1.60 to $2.40 per chrysalis—depending on supply and demand—the mariposario brings in enough income to significantly help families who raise butterflies and gather chrysalises on their own land. The income from these activities has lifted hundreds of Costa Ricans from poverty to the middle class.

After we've walked through the greenhouse garden, Vanessa shows me the pupae room, where long rows of chrysalises quietly dissolve and reorganize into butterflies. Some are inconspicuous, and would be almost impossible to find in the wild: The band-celled sister *(Adelpha fessonia)*, which will become a black butterfly with dramatic white stripes across its wings, looks like a shriveled brown leaf at this stage.

Others are showier. The familiar orange-and-black monarch *(Danaus plexippus)* completes its metamorphosis inside a pale aqua capsule encircled by a glinting golden ring. And the entire chrysalis of the orange-spotted tiger clearwing *(Mechanitis polymnia)* gleams in stunning gold.

"What about these metallic chrysalises? Don't they attract predators?" I'm not just thinking of hungry animals; these jewelry-like pendants surely attract humans, too. They look like actual gold, and I wouldn't mind hanging one on a thin chain around my neck, except that it's a living being, even in this sealed and silent state.

"We don't really know why some are metallic," Vanessa admits. "There's still a lot of research to be done.

The tour continues to the Gallery of Human Transformation, a room adorned with banners linking butterfly metamorphosis to human growth—terms like "supraconscious potentialities" and "eternal union with the Supreme" float above. I smile, enjoying the metaphor without needing the mysticism.

The tour is over now, but I cannot leave; I need to spend more time with the butterflies. The shady waterfall area—the place that's filled with blue morphos—is my favorite spot. It's at the far end of a quiet path, a private place for contemplation in the midst of kinetic splendor.

Later, it's closing time. The sun is slanting, and I must leave the mariposario. The visit has filled me with existential questions: Does a caterpillar have an inkling of what it will become? Does a butterfly remember its caterpillar days? The chrysalises captivate me most—unborn butterflies suspended in transformation, a tangible reminder of change and possibility. They mirror the spiritual and emotional work I'm doing in my own life: dissolving old patterns, reorganizing, and emerging renewed.

As I walk back to central La Guácima, I notice a mural I hadn't seen before. It depicts butterflies floating through space on ethereal wisps that remind me of the Butterfly Effect hypothesis: the idea that a slight change in one place—like the flapping of a butterfly's wings—can lead to big changes in another place or time. The butterflies and their chrysalises become talismans—not of my own transformation, but of a larger truth I've been watching all day: that change is incremental, and already underway.

Hiking through the Trinidadian jungle

Call of the Nightjar

On monocultures, white powder,
and enlightenment after dark

Miles from civilization, I sit on a scratched-up plastic chair, swatting at the insects buzzing around my face. The night sky glows, its Milky Way cutting a sparkling swath through the darkness. Cicadas hum and tree frogs trill. "Listen!" Nigel points into the forest. The call—deep, liquid, and jarring—echoes through the trees: *chuck-weep-wit-WEE-oh, chuck-weep-wit-WEE-oh.*

Nigel is a Trinidadian biologist educated in the U.S. His research has spanned primates, wildlife management, and ecological studies, but his heart is here, in his home rainforest. I've volunteered on an Earthwatch program about ocelots he co-directs—but tonight, the jungle itself will be the lesson.

I turn my head and peer into the darkness. "I've

been wanting to see one!" I say, recognizing the call of the nightjar. It is an odd nocturnal bird. A plain-looking, brown-and-white ground dweller, it is named for its "jarring" call, which I'd been listening to for hours every night since I'd arrived. In fact, it had been keeping me awake. But I couldn't find the bird during the day.

"You pass right by them every morning," Nigel says. "They're easy to find, if you know where to look. Maybe I can show you one tomorrow."

After a long day cataloging plants and camera-trapping ocelot trails, Nigel and I are relaxing at camp. He tells me about working with the pharmaceutical industry, the challenges of supporting tropical research, and the surprises hidden in Trinidad's streams.

"So, you support yourself doing research projects?" I ask.

Nigel laughs.

He pauses for a long time.

Finally he looks down and says, "I run."

"You run?"

"For a living," he says. "To support myself and my wife and child, who live back in the States." Nigel, who has decided that pharmaceutical-backed research is not a good fit for him, supports his family by extreme running—an endurance sport tougher than ultra-marathoning, with an emphasis on almost unimaginable distances, temperatures, and elevations —and treacherous terrain. He used to win enough prize money from extreme running to support his research. "But now, in my forties, I'm competing with people young enough to be my children." Nigel laughs again, ruefully. "It isn't sustainable."

Sustainability, I will soon learn, is Nigel's obsession.

Trinidad is home to the ocelots Nigel is studying, as well as a wide variety of tropical animals—monkeys, leatherback turtles, boa constrictors, anteaters, and agouti—and more than 400 species of birds. The island is a kaleidoscope of biodiversity, which is reason enough to love it, but there's something very special about the streams of Trinidad's northern mountain ranges, especially if you're a biologist.

"Guppies are an incredible resource!" Nigel says.

Surely I misunderstood him. These quickly reproducing aquarium fish sold for ten cents each at the pet store when I was a kid; they can't possibly be of any significant value. But it turns out that Nigel knows a lot more about guppies—and value—than I do.

He explains that the streams here, similar yet geographically isolated from each other, have created a unique natural laboratory. The guppies living in them have been studied and documented for many generations, giving scientists an unparalleled ability to learn about the ways evolution and ecology are intertwined.

"These streams have been epicenters for breakthroughs in evolutionary theory," Nigel says. "For example, some guppies evolve to be especially swift swimmers, like sprinters. But not all of them do, even though it makes sense evolutionarily for them to be fast, to escape predators."

"OK; why is that?" I ask, swatting at persistent mosquitoes and feeling a bit like a predator myself.

"That's what we're learning. The unique combination of isolation, species diversity, and system-replication has allowed scientists to learn a lot about cognition, cooperation, and lifespan."

"That sounds like valuable information ..."

"Yes, and that's what it's all about—*value assignment!*" Nigel interrupts me, nearly jumping out of his chair. I've stumbled onto his favorite topic. "Data from Trinidad's streams has powerful implications for models of evolution and conservation, as well as population management. So the northern range of Trinidad has an extraordinarily high global scientific value."

Nigel is educating me eagerly, his words rushing and tumbling like the waterfalls in those mountain streams. He is fascinated with the way people—especially people in power—assign value to various things, from scientific data to food and medicine. This obsession is bigger than his current research, bigger than the rainforest. It's about the very core of sustainable life on Earth.

"Value assignment tends to have cultural bias." Nigel leans forward. "It's even a problem for local scientists. The best and brightest go to industrialized countries for their education, and don't learn *tropical* environmental management. If they return to Trinidad, it's often with values reflecting the cultural biases of their white, First-World instructors—including monoculture agriculture."

"What? How is monoculture agriculture a bias?" I grew up in southern Iowa, with its rolling fields of corn and soybeans. The family farms there are not a bias; they're a way to make a living—and a time-honored tradition.

"We think of farmers as people who till the land," he continues. "But think about it: We can get only a tenth of the edible produce from monoculture, compared with a mixed-use agri-forest system. Which, by the way, describes the backyards of a lot of people living in the tropics."

"But what about a coffee or banana or cacao plantation," I frown. "Aren't they efficient enough to provide a lot of income?"

"There's a huge demand for caffeine, in the form of

coffee, and organizations like the Rainforest Alliance have tried to make coffee agriculture much more sustainable by verifying operations and informing consumers. But these are short-term benefits; there are more nuanced ways of looking at it."

Nigel continues, eyes shining. "Another perspective is that when you're marketing coffee—little caffeine packets—it's legalized drug trafficking. People can't subsist on coffee. Even the people who grow it can't subsist on it. They have to export it. They're locked into a specific economic package, which is dependent on the whims and the addictions of people who are wealthy enough to buy it. Is this really a sustainable practice? We have to think longer-term."

"Nigel," I interrupt, "You're giving me a real education."

"No," he insists. "I'm just reminding you of something you already know. People in the First World know this. They realize it when they're pushing paper at their desks, or sitting in traffic..." Nigel trails off, perhaps thinking about the traffic in Berkeley, where he earned his PhD. Then he starts up again.

"It's important to pay attention to what you already know. In marketing we literally create jobs for people to move a perception around. And those are the jobs that make the most money right now—those are the most valuable jobs. Those are the jobs people seek."

Nigel is right; this is information I already have. But now I'm thinking about it in a new way. I shift in my seat. We don't know each other very well. This doesn't seem like the right time to mention that I supported myself and my son for many years with one of those marketing jobs—but I am definitely thinking about it.

Perhaps Nigel senses my discomfort. He leans forward in his chair and changes the subject. "A lot of the perceived value of the rainforest has to do with the potential revenue for big discoveries. You know that the white powders all came from the rainforest."

"White powders?"

"Yes: sugar, arrowroot, cocaine. Did you know it was legal to buy heroin from Sears Roebuck at the beginning of the twentieth century, and that LSD research was once part of leading-edge cognitive science? There are stigmas attached to them now, but over time we change our attitudes about various chemicals, about certain drugs. Sugar, caffeine, and theobromine—that's chocolate—they're all legally marketed now."

"But what is sustainable in the long term?" Nigel continues. "Cacao is native to South America and now grows in Africa and around the world. Is it sustainable to expand cacao in Western Africa at the exclusion of native species? People can *look* at a cacao forest and see that it's less diverse than a rainforest."

Nigel stands and begins pacing, his long arms wheeling in the night air. "With a monoculture we're assigning value based on one product or one crop, and the decisions we make are based on the global marketplace, not a local marketplace. The result is increased output—but in a non-sustainable way: we move water around, we use pesticides, we poison the earth. So we're decreasing production and increasing management of temporary market value to make a few people very wealthy.

"Meanwhile, local people can't afford to feed themselves; it's cheaper for them to buy grapes at the supermarket than to grow their own. This is how it works in the global marketplace, where the drive is to *market* something. So you will forfeit subsistence to commute to a place where you can move this valueless information, which sequesters wealth."

I squirm again, my damp legs sticking to the plastic chair. Sequestering wealth, drug trafficking, benefiting from valueless information—suddenly I feel like one of the bad guys.

Nigel switches back to guppies. "Studying the ecology of wild guppies has led to insights about health care, behavioral models, and the design of cooperative networks."

Talking with him is like *wrangling* guppies; ideas dart every which-way, twisting around on themselves and diving off again in a new direction. "These applications are assigned high values in the global marketplace," he continues.

"So there's a high value in terms of both science and commerce! That's a good thing; value assignment is working the way it should," I venture.

"Yes, but the rainforest is disappearing because of the value assessment of businesses and the voting public in Trinidad. They think the jungle is only good for recreation, tourism, and cheap resource extraction."

I think I understand what Nigel is talking about. This is one of the things Earthwatch notes as a bright surprise about their programs: Having citizen scientists come from around the world to study a place often results in the local people valuing things they had previously taken for granted.

"There are so many treasures hidden here," I begin, as the nightjar calls again. "There's a whole medicine cabinet in the jungle. Maybe there's even a cure for cancer. If we would only appreciate what we have and preserve it..."

Nigel interrupts, choosing his words carefully: "The potential to discover cancer-curing drugs is not the only reason—or even the best reason—to preserve the rainforest," he explains haltingly, pausing to let his point sink in. He sits down again and looks out into the night—not at me. "That's a white, First-World reason."

"What do you mean, *It's a white, First-World reason?*" It seems like Nigel is accusing me of something, but I'm not sure what. I feel vaguely guilty. The conversation is awkward for both of us.

"The rainforest can have much more value than just being a grab bag for potential cures for cancer," Nigel explains. "Cancer is a minor problem for the wealthiest of the wealthy."

I have no idea what he's talking about. My mother and both my grandmothers died of cancer. It was a pretty big deal to us, and we're certainly not the wealthiest of the wealthy.

"In most of the world, people don't even live long enough to get cancer," Nigel continues. "What they need is food. The rainforest is important because it's a tropical breadbasket." His voice rises in excitement, and the jungle goes silent.

So we're all running around looking for miracle cures, I think, *and ignoring what's right in front of our faces—all the food the rainforest provides. It's the biodiversity that makes living here sustainable. And once we've lost it to monoculture ...* "How would we assign value to biodiversity?"

"We assign value only in terms of money, but we can expand how we think about it."

What Nigel's saying makes sense. Food is valuable, but so is scientific research. We may not realize its value for generations, though, and we can't put a number on it.

"What is the value of figuring out why some guppies become sprinters?" Nigel asks. "Or of knowing what determines their life span—or yours? Or finding a cure for cancer? There's so much we haven't discovered!"

I look at him quizzically. *Does Nigel realize what he just said?* Then we both burst into laughter.

"OK; I admit it," he chuckles. "I had to throw in *cure for cancer*."

We gaze up at the night sky as I contemplate my new understanding of the value of the rainforest. Nigel has led me into a maze as big as the forest itself: sustainability, the ecology of wild guppies, marketing and value assignment, cultural bias, biodiversity, curing cancer.

Chuck-weep-wit-WEE-oh. The nightjar calls again, a reminder that the forest exists for itself. The ecosystem feels alive in a way that no textbook or lecture could capture. Its worth can be measured not only in products or profits, but in survival, connection, and even awe.

In Gratitude

The animals in these pages gave me far more than memorable stories. They gave me moments of humility, wonder, and a deeper understanding of our shared vulnerability on this planet.

In gratitude, I am donating 100% of my paperback proceeds from this book's 2026 sales to wildlife advocacy organizations.

Supporting wildlife and their habitat feels, to me, like the most meaningful ending I can offer.

— Laurie McAndish King

Bonus Material

Thanks for reading! I hope these stories will encourage conversation and help promote wildlife conservation.

I fell in love with all the locations I visited, but Namibia was extra-special: It was the first country to include conservation in its constitution, and their broad-based efforts are paying off.

If you'd like to get a closer look at Namibia's desert-adapted rhinos, cheetahs, lions, and elephants, check out "Wild Namibia," a full-color PDF with more of my photos, and wildlife notes that I couldn't quite fit into this book.

Visit LaurieKing.com for:

- Book Club Discussion Guide
- "Wild Namibia" color PDF

Resources

If you'd like to know more about the animals and organizations in the book, these URLs are good as of press time:

- Save the Rhino: Savetherhino.org
- Earthwatch.org
- Centre ValBio/Ranomafana National Park (lemurs): https://www.stonybrook.edu/commcms/centre-valbio/visitcvb/visitingcvb/rnfPark.php
- Zealandia: VisitZealandia.com
- Cheetah Conservation Fund: Cheetah.org
- Fortress of the Bear: Fortressofthebear.org
- Asa Wright Nature Center: Asawright.org
- The Butterfly Farm: Costarica.com/attractions/butterfly-farm

Acknowledgments

A huge thank-you to Jim Shubin, who accompanied me on many of these trips, heard and re-lived the stories countless times, and designed this beautiful book; Linda Watanabe McFerrin, who edited most of the stories; Joanna Biggar, who convinced me to visit Namibia with her; Gil Zeimer for his close read; and Kaye McKinzie and Andréa Guevara, who each helped me navigate the vagaries of the publishing world. I also appreciate the opportunity to learn from Joy Williams at *Orion* magazine's Environmental Writers' Workshop.

I would be remiss not to mention the super-hero authors who inspired this collection. I don't know them, but I love their work: David Abram's *Becoming Animal*, Bill Bryson's *The Body*, Tim Cahill's *Jaguars Ripped My Flesh*, Robin Wall Kimmerer's *Braiding Sweetgrass*, Aimee Nezhukumatathil's *World of Wonders*, Katherine Rundell's *Vanishing Treasures*, Susan Orlean's *On Animals,* Michael Pollan's *The Botany of Desire*, David Quammen's *The Song of the Dodo*, Matt Ridley's *The Red Queen*, Mary Roach's *Replaceable You*, Lyall Watson's *Supernature*, E.O. Wilson's *Consilience*, Ed Yong's *An Immense World*—these books are among my favorites, and each one amplified my awe for the natural world. Reading them helped shape *Wild Thing, I Think I Love You!*

About the Author

Laurie McAndish King grew up in rural Iowa, studied pre-med and philosophy at Cornell College, and earned her master's degree in education. She observes with an eye for natural science, and writes with a philosopher's heart.

King's award-winning travel essays and photography have appeared in many publications, including *Smithsonian* magazine, the *Washington Post*, Lonely Planet's *The Kindness of Strangers* (with a preface by His Holiness the Dalai Lama), and more than a dozen literary anthologies.

Her writing won a gold Lowell Thomas award and her mobile app about the San Francisco Waterfront earned a 5-star rating on iTunes. King's audio work has appeared on KUSF, FCCFREERADIO, and NPR-affiliate KCBX.

Laurie lives in northern California and is an avid photographer (one of her photos was displayed for six months at the Smithsonian). She also enjoys travel, gardening, and falling down rabbit holes.

Connect at LaurieKing.com

Other Books by Laurie McAndish King

The Curious Traveler stories
—insightful, poignant and often quite funny—
are celebrated for the way they inspire and entertain readers.

Lost, Kidnapped, Eaten Alive!
True Stories from a Curious Traveler
"King brings uncommon heart and soul to her travel stories ... a rollicking ride."
—Michael Shapiro, Author of *A Sense of Place: Great Travel Writers Talk About Their Craft, Lives, and Inspiration*

Your Crocodile has Arrived:
More True Stories from a Curious Traveler
"King's curiosity is contagious ... Her tales glow with wit, warmth, and wonder."
—Jeff Greenwald, *Author of The Size of the World* and *Shopping for Buddhas*

An Elephant Ate My Arm:
More True Stories from a Curious Traveler
"Insightful tales that combine a love of nature with astute observations about the human condition ... a perfect read for book clubs."
—Judith Horstman, Author of *The Scientific American Healthy Aging Brain*

www.ingramcontent.com/pod-product-compliance
Lightning Source LLC
LaVergne TN
LVHW010339180926
848251LV00034B/239
9781732774766